Reinventing Ourselves

How Technology is Rapidly and Radically Transforming Humanity

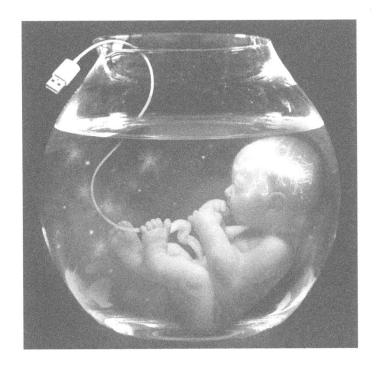

Larry Bell

Reinventing Ourselves: How Technology is Rapidly and Radically Transforming Humanity

© 2018 Larry Bell All Rights Reserved
Print ISBN 978-1-949267-14-3
ebook ISBN 978-1-949267-15-0

Other books by Larry Bell:

Scared Witless: Prophets and Profits of Climate Doom
Climate of Corruption: Politics and Power behind the Global Warming Hoax
Cosmic Musings: Contemplating Life beyond Self
Reflections on Oceans and Puddles: One Hundred Reasons to be Enthusiastic, Grateful and Hopeful
Thinking Whole: Rejecting Half-Witted Left & Right Brain Limitations

This book is sold subject to the condition that it shall not, by way of trade or otherwise, be lent, resold, hired out or otherwise circulated without the publisher's prior consent in any form of binding or cover other than that in which it is published and without a similar condition including this condition being imposed on the subsequent purchaser.

STAIRWAY PRESS—Apache Junction

Book Cover Art: Tamalee Basu
Cover Design by Chris Benson
www.BensonCreative.com

STAIRWAY≡PRESS

www.StairwayPress.com
1000 West Apache Trail, Suite 126
Apache Junction, AZ 85120

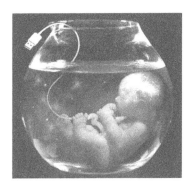

Dedication

Dedicated to optimists who rescue us from futility, pragmatists who warn us of perils, and idealists who elevate our purposes.

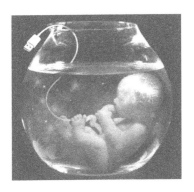

Introduction

A GROUP OF scientists, mathematicians and engineers from many organizations held a first meeting of a series at Dartmouth College in 1956 to consider a rather preposterous idea. They wondered if it might be possible to create "thinking machines" that could duplicate or surpass human intellectual capacities.

Two years later, Jack St. Clair Kilby, an electrical engineer at Texas Instruments, came up with another radical idea. He stayed in the lab during a company vacation break and cobbled together a calculating device that combined a transistor, a capacitor and three resisters on a single piece of germanium—the first integrated circuit. It wasn't particularly small—about a half-inch long but narrow—and not very elegant either. With wires sticking out, it resembled an upside-down cockroach glued to a glass slide.

In January 1959, Bob Noyce at Fairchild Semiconductor in Palo Alto, California developed a precision photographic printing technique using glass as insulation to deposit tiny aluminum wires above silicon transistors without the messy cockroach legs. Kilby's integrated circuit was transformed into a rapidly producible integrated chip with wires twenty times smaller.

The following year, Texas Instruments introduced a Type 502 Flip Flop with one bit of memory that it sold for $450. Weeks later, Fairchild produced its own model which was used by other computer companies, the U.S. Air Force and NASA's Apollo rockets.

One bit soon grew to 4, then to 16, then to 64. This capacity increase occurred as the chips continued to rapidly shrink in size.

According to a 1965 prediction by Fairchild research director Gordon Moore, now referred to as "Moore's Law," the chips' information density would double every 18 to 24 months. By 1969, the TI 3101 64-bit memory chip was priced at $1 a bit. Your iPhone probably has a trillion bits priced at merely picocents each.

Beyond that, the computers they contained are rapidly becoming smarter and are competing with practical roles we have naturally assumed needed us. Few occupations, including a major percent of highly-trained legal and medical diagnostic services, are immune from competitive AI and concomitant robotic automation workplace challenges.

Hedge funds are using AI to beat the stock market, Google is using it to diagnose heart disease more quickly and accurately and American Express is deploying bots to serve its customers online.

A 2016 McKinsey Global Institute study estimated that between 10 and 50 percent of all U.S. job tasks could be automated using existing robotic technology. In about 60 percent of 800 occupations surveyed, at least 30 percent of those primary activities can be replaced by software.

Some jobs, such as driving and working in retail and fast-food, may become entirely obsolete.[1]

Automated servants have already come into our homes to perform such domestic tasks as cleaning swimming pools, mowing lawns and cleaning floors. Others provide us with patient and informed responses to questions regarding the best traffic route to take to virtually anywhere and a limitless variety of other topics.

Don't be surprised to see interactive human-mimicking next-generation versions of Amazon's "Alexa," Microsoft's "Cortana," Apple's "Siri," and Google's "Google Assistant" increasingly serving as virtual substitutes for actual two- and four-legged companions. Many can already anticipate what you may wish to discuss based upon your past information requests. Within a decade, their conversations and voices will become indistinguishable from fellow humans who can be counted upon to tell you what you really want to hear.

Global populations, Americans included, are trading away more and more of their personal privacy for promises of increased convenience and security. Spy cameras are sprouting up on lampposts and rooftops everywhere, facial recognition systems can track each of our individual movements and Internet-connected "smart cities" are wiring private home appliances within municipal energy monitoring and eventual control networks.

A smart city goal is premised upon supporting better decisions about design, policy and technology on information from an extensive network of sensors that gather data on everything from air quality to noise levels to people's activities.

Some plans call for all vehicles to be autonomous and shared. Robots will then roam underground doing menial chores like delivering mail. Many of the same large American tech companies that will provide these devices are working with China to establish widespread public monitoring and social media censorship programs there.

Meanwhile, computer processing capacities are accelerating at an astounding rate with the advent of recent quantum computing (QC) advancements.

As University of Maryland researcher Christopher Moore testified at an October 2017 House Science Committee hearing on "American Leadership in Quantum Technology," merely 300 atoms under full quantum computer control might potentially store more pieces of information than the number of atoms that exist in the entire Universe.

QC capabilities offer potentials to understand much more about natural systems and phenomena which are ridiculously hard to model with a classical computer. For example, trying to simulate the behavior of the electrons in even a relatively simple molecule is an enormously complex endeavor.

In any case, there is no way to turn back the clock of progress where even Einstein's space-time continuum takes on a new dimension of meaning. Unlike the speed of light, there are no known theoretical limits to computational intelligence.

Whereas the computational power of the human brain is largely prewired by evolution, AI capacities arising from revolutionary computer technology power and applications are growing exponentially.

As Christof Koch, chief scientist and president of the Allen Institute of Brain Science in Seattle, predicts, sweeping societal and economic influences of AI signal a fourth industrial revolution.

He observes:

> The first, powered by the steam engine, moved us from agriculture to urban societies. The second, powered by electricity, ushered in mass production and created consumer culture. The third, centered on computers and the Internet, shifted the economy from manufacturing into services.[ii]

Koch points out that before modern farm equipment and tractors came along, it took thirty-times more people to farm one hundred acres than it does today.

This has resulted in producing more food for growing populations at affordable prices.

And while the Model T turned out the lights on many professions including blacksmiths and carriage makers, its introduction of affordable automobiles through mass production created huge new demands for labor created by the steel, glass, rubber, textile trade and oil and gas industries.

Dystopian visions of massive AI-driven job losses are premature. Throughout history, employment adapted as machines gradually replaced more and more aspects of labor over time. While once again, this fourth revolution will eliminate some jobs, it will also create opportunities for new ones that will require and enable more people to think smarter.

This book began with no doomsday preconceptions regarding AI, nor does it conclude with one. In any case, it is a natural tendency to impose bias that assumes that tomorrow will resemble our positive and negative experiences today, only a little bit different—or maybe extremely so—instead of recognizing that we are in the middle of an unknowably disruptive change.

Understandably, and appropriately, many people worry that AI will have innumerable and far-reaching disruptive effects on society—putting people out of work, adding to work and income inequality and some will say, even pose an existential risk to the long-term future of Homo sapiens.

Others will remind us that just as all of the previous technical revolutions profoundly increased human productivity, welfare and lifespans, AI will make human society better.

Both views will likely prove correct.

[i] *Where machines could replace humans—and where they can't (yet)*, Michael Chui and James Manyika, McKinsey Quarterly, July 2016.
[ii] https://alleninstitute.org/what-we-do/brain-science/about/team/staff-profiles/christof-koch/

Artificial Intelligence: Are We Outsmarting Ourselves? 4

Thinking Machines? 5

Multiple AI Evolutions and Revolutions 7

A Very Brief Early AI History 8

Breaking Moore's Law 10

AI that Mimics Human Learning and Behaviors 12

Quantum Computing Leaps Forward 14

Artificial Intelligence vs. the "Real Thing" 16

Getting Connected to the Cloud 20

Personal Connections To The Internet: Changing Our Minds And Relationships 25

Influences on How We Relate to One Another 29

Our Sense of Privacy and Security 34

Influences on Values and Behaviors 42

Dangerously Deceptive Virtual Friendships 45

Unnatural Selection in Mating Evolution? 47

Information vs. Learning and Thinking 50

Is Information Technology Rewiring Our Brains? 54

AI Destructions In The Workplace: Human Advancements Vs. Obsolescence 62

The AI Revolution 63

AI Influences on Creative Destruction 69

The Most AI-vulnerable Jobs and Careers 80

A Second Machine Age 84

AI Impacts on Healthcare 86

AI Impacts on the Practice of Law 90

Adapting to Disruptive AI Influences 92

Mechanizing Our Lifestyles: Utopian Resorts or Ant Farms? 97

Influencing Where and How We Live 98

Just How Smart are Smart Cities? 104

Trading Our Privacy for Convenience? 110

Will Automated Vehicles Kill Car Romance? 117

Delegating Business Decisions to Algorithms 123

Who Rules the Technology? 127

A New Evolutionary Era?: Influences Upon Culture and Values 139

Information Technology Masters and Martinets 141

Trading on Good and Bad Faith 145

A New Normal: Living With Cybersecurity Threats 155

Are Social Relationships the New Virtual Reality? 159

Machine Ethics as a New Religion? 162

Bioengineering our Human Society 167

Are We Ready for Posthumanism? 172

Human Evolution And Revolution: Trends, Triumphs and Trepidations 178

A Mixed Bag of Prophases 181

Will Tech Overlords Lord Over All? 183

Reinventing Ourselves

Cyber Attacks on Privacy and Security 187

Smart Technology Lapdogs and Luddites 191

Lifestyle Matters 194

Capitalism Obituaries are Premature 201

Some Grossly Speculative Scenarios 209

Smart City Benefits Prove Costly 210

Capitalism Flees to the Countryside 212

Endnotes **216**

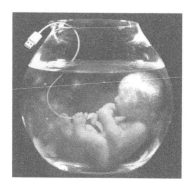

Artificial Intelligence: Are We Outsmarting Ourselves?

WILL THE RAPID computing evolutions ultimately revolt against us as HAL did against the Discovery One astronaut crew in the movie *2001: A Space Odyssey* based upon Arthur C. Clarke's novel series?

As you may recall, HAL 9000 (the Heuristically programmed Algorithmic computer) decided to kill the astronauts by locking them outside the spacecraft during a repair after reading their lips to discover a secret plan to disconnect the robotic system's cognitive circuits following lack of trust triggered by a computing glitch. HAL also attempts to suffocate hibernating onboard crewmembers by disconnecting their life support systems.

Fortunately, in this case, human dexterity wins the day. One of the astronauts circumvents HAL's control by manually opening an emergency airlock, detaching a door via its explosive bolts, reentering Discovery and quickly re-pressurizing the airlock.

Whew!

Nevertheless, can we truly blame HAL for lack of empathy or gratitude for its mentally inferior human creators? If they were smarter, wouldn't HAL's programmers recognize the risk that artificial intelligence might inevitably surpass the capabilities of our human "general intelligence"?

And after all, what characteristics will characterize "real" intelligence versus the evolution of artificial versions? Will computers always continue to depend upon information systems and data we teach them? Actually, some already teach themselves, and in the process, learn much faster than we do.

Thinking Machines?

The concept of "thinking machines," which was first hypothesized during a 1956 meeting of scientists, mathematicians and engineers at Dartmouth College, is no longer theoretical. Moreover, those machines are already beginning to outthink some top human experts in certain very complicated mental challenges.

- By 1997, a "Deep Blue" IBM computer defeated the reigning world chess champion, Gary Kasparov.

- In 2011, "Watson," another IBM computer, beat all humans in the quiz show *Jeopardy*.

- In 2016, an "AlphaGo" algorithm developed by "DeepMind," a London AI company, dispatched Lee Sedol, a top player in the ancient and complex board game "Go." The algorithm was originally trained on 160,000 games from a database of previously-played games.

The program was later upgraded to "AlphaGo Zero," which taught itself by playing four million games against itself entirely by trial and error. AlphaGo Zero subsequently annihilated its parent, AlphaGo, 100 games to zero. What it learned in less than one month would have required a decade or two of training for a human.

- In 2017, "Libratus" software developed at Carnegie Mellon University beat four top players over a 20-day tournament of No-Limit Texas Hold'em poker. The code doesn't need to bluff...it just out-thinks humans.[1]

We have witnessed a worldwide impact of artificial intelligence over just the last few years to the point that it dominates nearly all businesses, investments and even ethical narratives. As a result, two opposing attitudes appear to have emerged: one believing that AI will beneficially augment humans; the other that it will diminish them. Most likely prospects hold that both predictions are true.

There is virtually no likelihood that the AI revolutionary march of encroachment upon the human domain of activities will lose momentum. As Professor Justin Zobel, head of the Department of Computing & Information Systems at the University of Melbourne, Australia observes:

> It is a truism that computing continues to change our world. It shapes how objects are designed, what information we receive, how and where we work, and who we meet and do business with. And computing changes our understanding of the world around us and the universe beyond.[2]

Zobel notes, for example, that while computers were initially used in weather forecasting as no more than an efficient way to assemble observations and do calculations, today our understanding of weather is almost entirely mediated by computational models.

Another example relates to sweeping influences upon biological sciences and commercialization. Zobel points out that whereas e-research was once done entirely in the lab (or in the wild) and then captured in a model, it now often begins in a predictive model, which then determines what might be explored in the real world.

Justin Zobel alerts us to a new AI reality that computers are not only influencing a reinvention of ourselves, but also rapidly evolving towards a capacity to reinvent themselves. Here, he urges us to recognize that computing influences upon human activities and society often characterized as "digital disruption" are also disrupting the very nature and future of digital computing.

Multiple AI Evolutions and Revolutions

Professor Zobel directs our attention to a phenomenon whereby AI evolution is simultaneously advancing on separate, diverse and revolutionary tracks. He writes:

> *Each wave of new computational technology has tended to lead to new kinds of systems, new ways of creating tools, new forms of data, and so on, which have often overturned their predecessors. What has seemed to be evolution is, in some ways, a series of revolutions.*[3]

Zobel emphasizes that this technological explosion of computing computational capacities and applications is more than a single chain process of innovation which has traditionally been a hallmark of other physical technologies that shape our world. He cites as a previous example a chain of inspiration from the waterwheel, to the steam engine to the internal combustion engine. Underlying this is a process of enablement...the industry of steam engine construction yielded the skills, materials and tools used in the construction of the first internal combustion engines.

A big and consequential difference here is that in computing, something richer is happening where new technologies emerge...often not only replacing their predecessors, but also enveloping them. In doing so, computing is creating platforms on which it reinvents itself, continuously reaching up to the next platform.

A Very Brief Early AI History

So now, let's briefly review some recent—very recent—history along with the evolutionary and revolutionary march forward.

For starters, this so-called "digital computer revolution" first emerged in the commonplace public lexicon to describe rapid micro-processing-based advancements which began to occur during the 1970s and 1980s. Prior to that time, the only contact most people had with computers was through utility bills, banks and payroll services, or computer-generated junk mail.

Modern microcomputer ancestors used early integrated circuit (microchip) technology which greatly the reduced size and cost of their mainframe predecessors. Then, after "chip-on-a-chip" was commercialized, the cost to manufacture a computer system again dropped dramatically. Arithmetic, logic, and control functions that had previously occupied several costly circuit boards became available in a single integrated circuit which made high-volume manufacture possible. Concurrently, solid state memory advancements eliminated bulky, costly and power-hungry, magnetic core systems used in previous generations.

Early advancements were generally created by independent

entities and led to the availability of cheap, fast computing, of affordable disk storage and of networking. Within only a decade, computers became common consumer goods for word processing and gaming. Computing and information storage were contained in personal standalone units.

Speaking at a 1977 World Future Society meeting, Digital Equipment Corporation CEO Ken Olsen famously said:

> *There is no reason for any individual to have a computer in his home.*

His company reached its peak in the late 1980s as the number two computer manufacturer in the U.S. with sales revenues of $14 billion. Company setbacks were largely blamed on Olsen's failure to anticipate or understand a burgeoning computer market and led to Digital's desperation sale to Compaq in 1998...followed by a purchase by Hewlett-Packard in 2002.[4]

Personal computers increasingly earned their place in private homes and businesses by the late 1980s. Families, for example, found "kitchen computers" convenient for storing easily retrievable disk-based recipe catalogs, medical databases for child care, financial records and encyclopedias for school work.

Although predicted to be commonplace before the end of the decade, computers still weren't powerful enough to match more optimistic visions. Due to limited memory capacities, they could not yet multitask, floppy disk-based storage was inadequate both in capacity and speed for multimedia and display graphics were blocky and blurry with jagged text.

Networking technologies created in university computer sciences departments soon led to substantial collaborative software improvements. The resulting emergence of an open-source information culture then spread throughout wide user communities which took advantage of—and also contributed to—common operating systems, programming languages, and tools.

It took another decade for computers to mature sufficiently for graphical user interfaces to serve broad, non-technical user markets which gave rise to the Internet. Equipment and user costs dropped dramatically as data catalogs became maintained online and accessed over the World Wide Web rather than stored on floppy disks or CD ROM.

The global digital traffic infrastructure Internet formed as networks became increasingly more uniform and interlinked. Simultaneous increases in computing power and falling data storage costs rapidly expanded world-wide service markets. The Internet, which was popularized for personal email chat and business forums, also became a growing exchange mechanism for computer data and codes.

The marvelous confluence of networking, capacity and storage began in the 1990s. In combination with an open-source culture of sharing both leading and drawing from the Internet, it still remains in the infancy of a yet unknown evolutionary creature. It is highly speculative bordering on pure fantasy to imagine what forms will emerge years, much less decades, in the future.

Breaking Moore's Law

On the basis of computer capacity alone, a prediction was made by American engineer and Intel co-founder Gordon Moore in 1965. He noted that the number of transistors per silicon chip had doubled every year and predicted that the growth would continue over the following decade. By his estimation at that time, microcircuits of 1975 would contain an astounding 65,000 components per chip.

By 1975, as the rate of growth began to slow, Moore revised his time frame to two years. This time his revised law was a bit pessimistic; over roughly 50 years from 1961, the number of transistors had doubled approximately every 18 months. Nevertheless, Moore's estimates were broadly accorded great importance as virtual law.

Moore had based his prediction upon a dramatic explosion in circuit complexity made possible by steadily shrinking sizes of transistors over the decades. Measured in millimeters in the late 1940s, the dimensions of a typical transistor in the early 2010s were more commonly expressed in tens of nanometers (a nanometer being one-billionth of a meter)—a reduction factor of over 100,000.

Transistor features measuring less than a micron (a micrometer, or one-millionth of a meter) were attained during the 1980s, when dynamic random-access memory (DRAM) chips began offering megabyte storage capacities.

At the dawn of the 21st century, these features approached 0.1 micron across, which allowed the manufacture of gigabyte memory chips and microprocessors that operate at gigahertz frequencies. Moore's law continued into the second decade of the 21st century with the introduction of three-dimensional transistors that were tens of nanometers in size.

Problems with Moore's original capacity estimate which surfaced in 1975 had encountered a technical snag due to limitations posed by the photolithography process used to transfer the chip patterns to the silicon wafers which used light with a 193 nanometer wavelength to create chips which feature just 14 nanometers. Although the oversized light wavelength was not an insurmountable problem, it added extra complexity and cost to the manufacturing process. And while it has long been hoped that extreme UV, with a 13.5nm wavelength, will ease this constraint, production-ready EUV technology has proven difficult to engineer.

A roadmap around chip limitations sometimes described as "More than Moore" applies highly integrated chips which combine a diverse array of sensors and low-power processors. With the growth of smartphones and the "Internet of Things", these processors include RAM, power regulation and analog capabilities essential for GPS, cellular, Wi-Fi radios and even advanced microelectromechanical components such as gyroscopes and accelerometers.

Computer chip technology advancements are applying new materials such as indium antimonide, indium gallium arsenide and carbon (both in nanotube and graphene forms) which promise higher switching speeds at much lower power than silicon. Coming soon, expect monolithic 3-D chips, where a single piece of silicon has multiple layers of components built upon a single die.

Something else to expect...Moore's innovative lawbreakers will continue to produce computing processors which are more versatile, faster, smaller and energy-efficient in response to ever-growing demands for smarter systems that support and compete with human enterprises.

AI that Mimics Human Learning and Behaviors

Will computers like the fictional HAL in *2001: A Space Odyssey* ever take on intellectual qualities often associated with human experience lessons and emotions? Actually, some already are.

A technique called "Generative Adversarial Networks" (GANs) trains competing AI algorithms to challenge each other, learn from mistakes, and even to fool one another with convincing deceptions.[5]

GANs is an example within a wide-ranging AI technology field broadly referred to as machine learning which essentially mimics the way we learn from trial and error using positive and negative reinforcement methods to achieve a desired outcome. The process uses two opposing reward or loss functions, one a generative model (also known as the environment), and the other a discriminative model (also known as an agent).

In one example, a GANs generator randomly creates images that the discriminator must identify as either real, or recognize to be an artificial fake. Both entities are trained over a large number of iterations, with each iteration improving the "skill" ability of each. Over time, the discriminator learns to ever-more reliably tell fake images from real images, while the generator uses the feedback from the discriminator to learn to produce more convincing fake images.

Another example of a multi-agent GAN is "Style Transfer," where the model is provided two photos and two discriminators are tasked to produce a single picture. One of the discriminators is rewarded by conserving the content of the first image, while the other is rewarded by preserving the style of the second image. In the case of one discriminator presenting an abstract pattern—and the other presenting a picture of an elephant—each will work hard to come up with a compromise which satisfies both. Applying loss/reward criteria, a single picture will result after perhaps millions of iterations.

Successful GANs applications for text still have a way to go. Jeremy Howard, a University of San Francisco faculty member, developed a bot which was trained to speak like Friedrich Nietzsche after being provided the complete writings of that author. After a large number of iterations, the generator started to speak in a manner similar to Nietzsche, but the sentences didn't make sense. While companies are working on language-to-meaning mapping, we are still a long way from being there.[6]

As for mimicking life-like voices, that's far more successful. The systems can learn to reproduce a given text string of realistic impersonations after being given about 20 minutes of voice samples.

Human-mimicking versions of Amazon's "Alexa," Apple's "Siri," Microsoft's "Cortana" and Google's "Google Assistant" already track our movements and can listen in wherever we are. Many can already anticipate what we may wish to discuss based upon past information requests. Very soon, their interactive conversations and voices will become indistinguishable from fellow humans.[7]

Do be advised, however, that we may be wise not to trust them to hold those friendly conversations confidential. In addition to being developed to retrieve information, arrange appointments, provide directions, play favorite music and remind us of scheduled appointments, they can also be used to whisper our personal secrets to others behind our backs.[8]

So maybe be a little careful when engaged in private telephone conversations. That "person" you think you are sharing secrets with might actually be a robotic HAL or HILLARY replica of your imagined friend with unfriendly intentions.[9]

Quantum Computing Leaps Forward

University of Maryland researcher Christopher Moore testified at an October 2017 House Science Committee hearing on "American Leadership in Quantum Technology," merely 300 atoms under full quantum computer control might store more pieces of information than the number of atoms that exist in the entire Universe.[10]

Such implausible features are made possible by equally incomprehensible subatomic-scale phenomena. Unlike current computers which process tiny "bits" of data in a linear sequence as either a one or a zero, at the seemingly weird subatomic scale, a quantum bit (or "qubit") can be both a zero and a one at the same time. As a result, rather than growing linearly, adding more qubits expands computing power exponentially.

Quantum systems are potentially capable of computational feats that have proven to be inconceivable with conventional technologies. For example, they might be used to model molecules which are ridiculously hard to model with a classical computer...trying to simulate the behavior of the electrons in even a relatively simple molecule which is enormously complex. IBM researchers used a quantum computer with seven qubits to model a small molecule made of three atoms.[11]

Writing in the *Wall Street Journal*, Committee for Justice President Curt Levy and Ryan Hagemann at the Niskanen Center posit a challenge in ensuring ways to ensure that future AI algorithms with minds of their own remain accountable to transparent oversight.

The authors' greatest concern isn't that advanced computers we create will go rogue and turn against us like HAL in *2001: A Space Odyssey*. They foresee a greater threat that AI complexity enables developers to secretly "rig" a system to the advantage of special interests, such as to manipulate a corporate operating program to reveal trade secrets to outside competitors.[12]

QC progress now continues to rapidly accelerate following a three-decade scientific siesta since the concept was first proposed by Russian mathematician Yuri Manin in 1980:

- D-Wave Systems, a company based in Burnaby, British Columbia demonstrated a special-function 16-qubit QC in 2007 at the Mountain View, California Computer History Museum.

- In 2011, D-Wave Systems sold its first 128-qubit commercial system ("D-Wave One") to the Lockheed Martin Quantum Computing Center located at the University of Southern California. The companies have since entered into multi-year agreements which have led to the development of more powerful D-Wave Two and D-Wave 2X systems.

- In 2013, Google established a Quantum Artificial Intelligence Laboratory (QAIL) at NASA's Ames Research Center at Moffett Field, California in collaboration with the Universities Space Research Association (USRA).

- In 2015, QAIL publicly displayed a 10-foot-tall D-Wave 2X unit chilled at 180 times colder than deep space which is expected to operate 100 million times faster than any conventional computer.

- IBM has recently announced an initiative to build a commercially-available "IBM Q" along with an Application Program Interface (API) to enable customers and programmers to begin building interfaces between the company's existing five-qubit cloud-based computer and conventional computers.

Artificial Intelligence vs. the "Real Thing"

Is artificial intelligence different from "the real thing" that goes on in our brains? Apart from our claims to unique spiritual, sensate and social qualities, can our human species keep up with—or perhaps even survive—exponentially growing artificial thinking machines and automated surrogates which are already outsmarting and outworking us in many areas of human endeavor?

Or on the other hand, might we not only learn from those products of our own innovation, but even interface AI systems with human minds to expand cognition and consciousness? Writing in the *Wall Street Journal*, Christof Koch believes that we can.

He urges:

There is one way to deal with this growing threat to

our way of life. Instead of limiting further research into AI, we should turn it into an exciting new direction. To keep up with the machines we're creating, we must move quickly to upgrade our own organic computing machines: We must create technologies to enhance the processing and learning capabilities of the human brain.[13]

Koch offers some early, yet promising neurotechnology examples that apply to present-day computational systems:

- Transcranial direct current stimulation:

This noninvasive brain technology induces a weak electric field in the cortex underlying the skull. Research in animals and humans suggests that this may enhance neuro-plasticity, the process in which the brain improves its performance when an action is repeated over and over. Users wear headphones that gently stimulate the motor cortex while performing simple activities such as lifting weights, swinging a golf club or playing a piano.

- Electroencephalogram (EEG):

Electrodes built into a headset detect brain waves during deep sleep. The device then plays low sounds that enhance the depth and strength of those waves, leading to more restful sleep.

This noninvasive technology is presently limited because those billions of tiny nerve cells that generate brain waves are quite remote from the scalp, allowing only faint echoes of neuronal chatter to be picked up. Christof Koch concludes:

We aren't anywhere close to selectively silencing or amplifying the activity of small cliques of neurons.

- Neurosurgical Implants:

Ultimately, to boost brain power we need to directly listen to and control individual neurons and atoms of perception, action, memory and consciousness. This currently requires some neurosurgery to penetrate the scull and access brain tissue. The good news is that brain-machine implant interfaces are happening faster than expected.

Nancy Smith was injured in a car accident which left her as a tetraplegic who can only move her shoulder and head. Neurosurgeons and neuroscientists implanted a tiny "bed of nails" in the region of her cortex to encode her intention to grasp a cup or press piano keys. Algorithms decode her neural signals and pass instruction to a musical synthesizer so that she can play music in her mind.

Bill Kochevar was paralyzed below the shoulders following a bicycle accident. A Cleveland-based team of doctors and neuroscientists placed electrodes into his left motor cortex that read out electrical tremors of about 100 neurons. From these they decoded and transmitted his intentions to reach out and grasp objects by electronically stimulating muscles in his arm. While crude, it enables Kochevar to eat and drink by himself.

There are more than fifty patients with such neuronal listening devices installed in their brains. Current and future applications include direct brain stimulation for obsessive-compulsive disorder, treatment-resistant depression, essential tremor, Parkinson's disease, epilepsy, stroke recovery and even blindness.

Christof Koch visualized that new neurotechnology developments may help patients recover lost functionality, including driving a car with their minds, plus a great deal more. He contemplates:

My hope is that someday, a person could visualize a

> concept—say, the US Constitution. An implant in his visual cortex would read this image, wirelessly access the relevant online Wikipedia page and then write its content back into the visual cortex, so that he can read the webpage with his mind's eye. All of this would happen at the speed of thought.[14]

Koch continues:

> Another implant could translate a vague thought into a precise and error-free piece of digital code, turning anyone into a programmer. People could set their brains to keep focus on a task for hours on end, or control the length or depth of their sleep at will.

Koch's vision brings a literal new meaning to the notions of "getting our heads together on an idea" and "sharing thoughts." As he imagines this, he suggests:

> Another exciting prospect is melding two or more brains into a single conscious mind by direct neuron-to-neuron links—similar to the corpus callosum, the bundle of two hundred million fibers that link the two cortical hemispheres of a person's brain. This entity could call upon the memories and skills of its member brains but would act as one 'group' consciousness, with a single, integrated purpose to coordinate highly complex activities across many bodies.

Christof Koch believes that humankind is at the threshold of a transformational new era that merges unlimited capacities of thinking machines and biological minds to revolutionize the entire meaning of "intelligence:"

> *While the 20th century was the century of physics—think of the atomic bomb, the laser, the transistor—the 21st will be the century of the brain—the most complex piece of highly excitable matter in the known Universe.*

Ultimately, our intelligent machines may even influence us to reinvent ourselves and equip us with "bigger brains" that will be needed to keep pace with our inventions.

Getting Connected to the Cloud

History has demonstrated that unexpected inventions can rapidly transport our brains, bodies and human potentials along new, unchartered and transformative pathways. As Henry Ford purportedly said: "If I had asked my customers what they wanted, they would have said a faster horse."

Cloud computing is more than just a replacement for faster horses. It is a technological stampede of AI applications which are multiplying and accelerating at warp speed.

As Joe Baguely, the chief of technology officer for VMware EMEA, observes:

> *Just as email rendered the memo obsolete, cloud computing is set to impact the way we do business, offering a competitive advantage to those organizations bold enough to think outside the accepted.*

Baguely adds:

> *The technology is no longer 'nice-to-have,' but is a critical part of any organization's infrastructure. We're seeing more and more businesses relying on*

> *mobile devices as the era of 'office working' slowly draws to a close.*[15]

Joy Tan, writing in Forbes.com, emphasizes that cloud computing is becoming especially transformative for companies, especially small and mid-sized businesses. This trend will accelerate as data analytics, AI and other capabilities become available across a digital ecosystem comprised of thousands of distinct and separate clouds. As user group examples, she predicts:

> *A commercial aviation cloud will help airlines manage ground operations such as maintenance, fueling, baggage handling, and cabin cleaning, thereby increasing efficiency and helping flights take off on time. A utilities cloud will automatically repair faults in the power grid to ensure that homes and businesses get the electricity they need. A banking cloud will let financial institutions scan thousands of transactions per second to prevent fraud.*[16]

A big advantage of cloud computing is that it speeds up and reduces the costs of existing processes through resource pooling which allows customers to flexibly access any level of service they require without investing separately in costly equipment, software and operational labor hours. This value is driven by a transition from using physical resources—to those of online services (infrastructure-as-a-service, platform-as-a-service, software-as-a-service) that deliver similar functions.

User subscribers pay only for what they use without a need to monitor usage by themselves. In the case of a public cloud, it is a shared server that can be located anywhere in the world where customers do not necessarily know where it is.

Although AI advancements are driving cloud applications

exponentially, the general concept of delivering shared computing services through a global network dates back about six decades.

In 1962, John McCarthy, a well-known computer scientist who is credited as one of the AI founders, prophesized: "Computation may someday be organized as a public utility."

Cloud is not a new idea...it goes back about six decades. In the early sixties there existed the image of delivering computing resources through a global network. J.C.R. Lickider pioneered the idea of "intergalactic computer network" in 1969. His aim was to develop an interconnected worldwide network which would enable access to data programs wherever and whenever needed.

Cloud computing rapidly evolved in several directions during the 1970s with the advent of Web 2.0., IBM's breakthrough VM (virtual machine) operating system which allowed multiple virtual systems on one physical device.

Fast forward to the 1990s, when Professor Ramnath Chellappa of Emory University and the University of South Carolina defined cloud as the new "computing paradigm where the boundaries of computing will be determined by economic rationale rather than by technical limits alone."

In 1999, Salesforce.com started up and practiced the concept of delivering "enterprise applications" via a simple website through the use of distributed computing power. This was an important early milestone in cloud history.

In 2002, Amazon launched the Amazon Web Services which played a leading role in the new cloud computing field. Then in August 2006, Amazon released its Elastic Computing Cloud (EC2), a commercial web service which allowed users to rent resizable virtual servers on their computer applications. This development made web-scale cloud computing easier for small companies and individual developers.

In 2009, computing companies led by Google began to

provide browser-based enterprise applications, such as Google App.

The story continues…the evolution has only begun.[17]

While not a new idea, true cloud capabilities are only now beginning to be realized. Revolutionary applications are endless, and include management of the digital infrastructure of tomorrow's cities, management and monitoring of driverless car and drone taxi networks and ensuring more efficient operations of farms and power plants.

Cloud will continue to support and empower emerging technologies such as AI and help them to adapt to new platforms such as mobile smartphones, which overtook sales of PCs in 2011. These devices capture lots of unstructured data such as emails, text messages and photos which require more data analysis, time and processing power than most smartphones have.

The solution is to divide the computational role between the cloud and the phones which take advantage of an AI process of "inference" which smartphones use constantly. This always-on intelligence enables the devices to respond immediately to voice commands, ensure that photos are cataloged according to content and set cameras perfectly for different subjects under different shooting conditions.

Since inference needs to process data in real-time, all the time, even tomorrow's super-advanced smartphones won't be able to meet computing demands imposed by AI. They'll continue to have to rely on the processing power of the cloud.

The same cloud dependency will apply to driverless cars and trucks. Such vehicles will be provided with sensors and cameras that generate huge amounts of data which must be processed in real-time…some within the vehicle, but also by the cloud.[18]

Streaming media technologies, which have improved significantly since first introduced in the 1990s, have enhanced cloud customer benefits and content provider revenues. Here, the video or audio content is sent in compressed form over the Internet and played immediately, rather than being saved to a hard drive. This means that a user doesn't have to wait to download a file to play it because the media is sent in a continuous stream of data that can be observed as it arrives.

Users can also pause, rewind or fast-forward the content just as they could with a downloaded file, provided that the content isn't being streamed live. In addition, streaming makes it possible for users to take advantage of interactive applications, such as video search and personalized playlists.

Streaming benefits content deliverers as well. It enables them to monitor what the visitors are watching and how long they are watching it. In addition, it provides content creators with more control over intellectual property because the video file is not stored on the viewer's computer. Once the video data is played, it is discarded by the media player.

Video streaming is already demonstrating dramatic disruptive influences upon traditional retail shopping businesses and consumer habits. As Bernard Golden, the CEO of Navica observes:

> *Once you've experienced convenience in one part of your life and can see that it's possible, say, to substitute a five minute online task for a lengthy errand, you begin to expect that in all aspects of your daily experience.*[19]

And as the adage goes: "we ain't seen nuthin' yet."

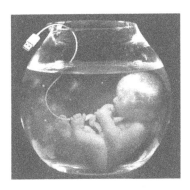

Personal Connections To The Internet: Changing Our Minds And Relationships

THE EMERGENCE OF IBM's Web 2.0 during the first decade of the twenty-first century fostered the rise of a series of Internet-based social and technical revolutions including social media phenomena and an endless variety interactive, multidisciplinary communication and information platforms.

The information revolution is bringing people of different backgrounds from around the world into a global information super network platform which connects many thousands or millions of social networks located everywhere.

Individuals are now empowered to create and exchange data with others from one end of the world to another in seconds; to make and share online presentations including videos, graphics, sound and text content; and to escape reality in virtual "game worlds." Suddenly, local issues and events captured on smartphone cameras became global; highly personal stories went public; and elections were won and lost based on social media campaigns.

Social media connections through the Internet have forever changed our senses of personal space, our life and work opportunities and countless aspects of our daily living routines.

They have altered the ways we interact with loved ones, friends, employers and clients; have nurtured and supported a proactive business startup culture enabled through electronic commerce; and have opened up ways to identify and actively participate with groups and individuals who share our special interests and problems.

E-Commerce has dramatically transformed the ways we acquire and provide goods and services. We can now book airline tickets and hotel reservations online; purchase an antique car part on eBay or an out-of-print book on Amazon; and even purchase fresh food with no need to travel to the grocery store. We don't have to stand in bank lines anymore either, and can pay for all purchase transactions using electronic credit cards and PayPal.

Since online shopping is removing the middleman, we can now purchase many of those products at much cheaper prices. Geographic distance is no longer a limit. The Internet has brought the power of online shopping and auctioning to people all over the world.

The Internet has freed us from geographic fetters and has connected us together in topic-based communities as a networked, globalized society connected by new technologies. It now provides much of our news and pundit views. It connects us with real-time updates about happenings around the world, including real-time sports and weather information.

We also have new means to broadly communicate our own views and creativity through blogs and books published through e-publishers. We can read them online or on Kindle. We advertise and transact our goods and services across oceans and deserts.

We now have just-in-time information about almost anything from anywhere. Print newspapers and magazines with dated information are being rapidly replaced by "news-breaking" electronic versions. Paperless record-keeping and publishing has replaced telephone books and cook books. Although not nearly as reliable, *Wikipedia* has replaced expensive and space-consuming sets of the *Encyclopedia Britanica*, which went out of print in 2011.

Many people turn to sites like AllRecipes and Food.com to decide what to eat and how to prepare it. On-demand music through Spotify, Pandora, iHeartRadio, and Slacker has eliminated a need to go to music stores, or to store physical recordings.

As many become more and more comfortable opening themselves up to opportunities online, they do so at risk to others who will exploit and misuse personal information. Online dating has become a major way to meet romantic partners...many are logging in to find love, rather than finding someone in their actual physical life who they can calibrate through direct sensate experience.

The Internet also enables us to learn about job opportunities and apply online. Again, this poses a potential problem of becoming stereotypically and narrowly "matrixed" through standardized human resource screening protocols which take personal contact out of the processes.

And with the advent of smartphones, we also no longer need wrist watches.

As blog writer Taryn Dentzel noted in an Openmind.com article, *How the Internet Has changed Everyday Life*:

> The future of social communications will be shaped by an "always-online" culture. Always online is already here and will set the trend going forward. Total connectivity, the Internet, you can take with you wherever you go, is growing unstoppably. There is no turning back from global digitization.[20]

Dentzel adds:

> The Internet has turned our existence upside down. It has revolutionized communications, to the extent that it is now our preferred medium for everyday communication. In almost everything we do, we use Internet. Ordering a pizza, buying a television, sharing a moment with a friend, sending a picture over instant messaging. Before the Internet, if you wanted to keep up with the news, you had to walk down to the newsstand when it opened in the morning and buy a local edition reporting what had happened the previous day. But today a click or two is enough to read your local paper and any news source from anywhere in the world, updated to the minute.

The rise of the Internet has sparked much discussion and debate regarding how online communication affects social relationships. Taryn Dentzel refers to the impacts of social media as having created a new "communication democracy" where the real value is that you stay in touch from moment to moment with people who matter most to you:

> Social media let you share experiences and information, they get people and ideas in touch instantly, without frontiers. Camaraderie, friendship, and solidarity—social phenomena that have been around us as long as humanity itself—have been freed from the conventional restrictions of space and time.[21]

The Internet has removed previous communication barriers. Online, the conventional constraints of space and time disappear and there is a dizzyingly wide range of communicative possibilities where we can share private gossip and jokes, participate in and watch online conferences, and join special groups to keep abreast of their specific interests.

Dentzel points out that the Internet is also bringing culture closer and more easily and quickly accessible to everyone. In doing so, it nurtures the rise of new forms of expression for art and the spread of knowledge whereby it has become "not just a technology, but "a cultural artifact in its own right."

This complex and transformative medium enables diverse cultures of thought and identity to flow freely across borders so that prior concepts of time, space and distance are losing conventional meanings.

Influences on How We Relate to One Another

A 2006 article published by the Pew Research Center asks, "Does the internet degrade friendship, kinship, civic involvement, and social capital?" Even in those relatively early Internet days, it posited that a great debate loomed regarding its positive and negative influences upon how Americans, in particular, relate to fellow friends, relatives, neighbors and workmates.[22]

The article points out that on one hand, many extol the Internet's abilities to extend our relationships—we can contact people across the ocean at the click of a mouse...we can communicate kind thoughts at two in the morning and not wake up our friends. Some prophets had predicted a decade earlier that the Internet would create a global village, transcending the boundaries of time and space.

On the other hand, others fear that the Internet causes a multitude of social and psychological problems which include a form of addiction similar to gambling. Nevertheless, unlike a recognition that one is "gambling too much," this is likely a poor comparison in terms of impact with the notion of someone "communicating too much."

The Pew Research Center authors, John B. Horrigan, Barry Wellman and Jerry Boase observe that a more pervasive concern among psychologists and social scientists is that the Internet "sucks people away from in-person contact, fostering alienation and real-world disconnection."

The underlying concern is that people become seduced into spending time online at their home computer screens at the expense of time spent with friends and family or visiting next-door neighbors. Critics worry that relationships that exist on text—or even screen-to-screen on flickering webcams—are less satisfying than those in which people can really see, hear, smell and touch each other.

The authors argue that debate regarding the Internet's impacts upon relationships is important for four basic reasons:

- There is a direct question regarding whether relationships in the Internet age will result in the same kinds of personal ties—in both quantity and quality—that flourished in the past: Do people have either fewer or more relationships? Is the ability to connect instantly around the world causing far-flung ties to dominate neighborly relationships? More broadly, does the Internet take away from people's overall in-person contacts, or add to them?

- There is an associated question of whether the Internet is splitting people into two separate worlds, one online—the other, offline. Both those who originally cheered and feared the Internet assumed that their "life on the screen" would remain quite separate and distinct from their existing "real life" relationships. Now it seems that this is not necessarily the case. For many, the Internet has become an integral part of numerous and varied ways that they relate to friends and neighbors. This begs a question, "Can online relationships be meaningful—perhaps even more meaningful than in-person relationships?"

- Another question revolves around whether people's relationships (both on and offline) provide usable help. Do they add to what social scientists refer to as "interpersonal social capital," such as providing useful information or emotional support, lending a cup of sugar or providing long-term healthcare? The authors point out that while it is easy to give information on the Internet and arrange for people to visit and help, it is impossible to change bedpans online. In contrast with the nostalgic way many people viewed small-town pre-internet social communities "where everybody knows your name," the authors ask: "where do they find community now?"

- Finally, the authors ask: "to what extent is the Internet associated with a transformation of American society from groups to networks?" Harking back again to mythological days for many when the average American had two parents, a single boss, and lived in that friendly village or neighborhood where everyone knew their names, evidence suggests that this traditional vision has changed.

North Americans are no longer bound up in a single neighborhood, friendship or kinship group. Instead, they tend to maneuver in social networks which consist of multiple and separate clusters. Where most friends may not know each other, and even more likely, neighbors don't know a person's friends or relatives.

Relationships have come to be widely spread across cities,

states and even continents. Accordingly, instead of a single community that provides a wide spectrum of help, relationships have become more specialized, where, for example, parents provide financial aid, and friends provide emotional support.

The authors conclude on a positive note, concluding that the Internet is not destroying relationships or causing people to be anti-social. Rather, it is enabling people to maintain and often strengthen existing ties as well as forge new ones.

As for time spent online reducing social contact, it may mostly draw away from time spent upon unsocial activities such as watching television and sleeping. The large amount of communication that takes place is often with the same sets of friends and family who are also contacted in person and by phone. This, the authors believe, is especially true for socially close relationships—the more close friends and family are seen in person, the more they are contacted online.

But even if many people tend to be even more connected with existing and expanded communities through the Internet, is the very nature of those connections altering the substance and quality of interpersonal relationships?

Writing in PsychologyToday.com, Alex Lickerman, M.D., urges us to remember that while we may enjoy online relationships using social media sites like Facebook or Twitter, for example, the differences between these kinds of interactions and those with people in the physical world are vast.[23]

Lickerman points out:

> *As long as we expect no more from these online relationships than we can give, no reason exists why we can't enjoy the power of social media sites to connect us efficiently to people we'd otherwise not touch. The problem, however, comes when we find ourselves substituting electronic relationships for physical ones, or mistaking our electronic*

> *relationships for physical ones. We may feel we're connecting effectively with others via the Internet, but too much electronic-relating paradoxically engenders a sense of social isolation.*

Electronic communication versus face-to-face contacts comes with other problems. For example, writing things like "LOL" and "LMAO" to describe laughter is no substitute for actually hearing and seeing someone laugh to lift our spirits when we're feeling low. Lickerman recommends making it a rule of thumb to limit email communications primarily to factual information, making it a priority to communicate emotionally sensitive or satisfying conversations more directly in person.

There can also sometimes be a tendency to use electronic media to avoid uncomfortable face-to-face topics and potential confrontations because it blocks us from registering negative emotional responses such messages engender. In these cases we intentionally isolate ourselves behind the medium's poor ability to transmit empathy.

Dr. Lickerman suggests that the "emotional invisibility" we witness on the Internet explains much of the vitriol we witness on many websites:

> *People clearly have a penchant for saying things in the electronic world they'd never say to people in person because the person to whom they're saying it isn't physically present to display their emotional reaction. It's as if the part of our nervous system that registers the feelings of others has been paralyzed or removed when we're communicating electronically, as if we're drunk and don't realize or don't care that our words are hurting others.*

Lickerman concludes that while the Internet is an amazing tool which has shrunk the world and brought us closer together, it also threatens to push us apart:

> Like any useful tool, to make technology serve us well requires the exercise of good judgment. For whatever reason, the restraints that stop most of us from blurting out things in public we know we shouldn't seem far weaker when our mode of communication is typing. Unfortunately, typed messages wound even more gravely, while electronic messages of remorse paradoxically have little power to heal.

He continues:

> Perhaps we just don't think such messages have the same power to harm as when we say them in person. Perhaps in the heat of the moment without another's physical presence to hold us back, we just don't care. Whatever the reason, it's clearly far easier for us to be meaner to one another online. Let's try not to be.

Our Sense of Privacy and Security

Those wonderful Internet social media platforms which enables us to locate, connect and interact with one another accordingly also pose real challenges which threaten our privacy and security. Writing in Openmind.com, Zaryn Dentzel warns:[24]

> Much of the time, people started to use social media with no real idea of the dangers, and have wised up only through trial and error—sheer accident, snafus, and mistakes. Lately, inappropriate use of social media seems to hit the headlines every day.

Celebrities posting inappropriate comments to their profiles, private pictures and tapes leaked to the Internet at large, companies displaying arrogance toward users, and even criminal activities involving private-data trafficking or social media exploitation.

Thanks to the Internet, a complete stranger from across the globe may view your Facebook profile and learn many personal things about you...gone are the days when privacy meant locking your front door.

Nude photos and other personal blunders refuse to die or recognize geographical barriers on the Internet. "Revenge porn" has become widespread. Nudes posted online, like drunk tweets and Facebook rants about an employer, can live on in infamy. New companies are being born that exist solely to clean up customers' online reputations.

According to a 2014 Pew Research Center report, technology in bonded relationships now encompasses even intimate moments. Sexting, or sending sexually suggestive nude or nearly nude photos and videos via cell phone, is practiced by couples and singles alike. In 2014, nine percent of adult cell owners had sent a "sext" of themselves to someone else, up from six percent who said this in 2012.[25]

The privacy issue is particularly sensitive to minors. Despite widespread attempts by schools, parents and concerned private and public organizations to raise their awareness, children continue to behave recklessly online.

Cyberbullying has become increasingly common, especially among teenagers. Emotionally-harmful behaviors include posting online rumors, threats, sexual remarks, a victim's personal information, unflattering recorded impersonations, body-shaming statements and images and pejorative labels.

Unlike traditional bullying, victims often may not know the identity of the bullies, or why they are being targeted. This can have wide-reaching effects on the victim because the content can be easily spread and shared among many people and remains accessible long after the initial incident. Also, since it often operates using stealth, the victim has no way of avoiding or escaping attacks.

Cyberbullying is also easily accomplished. It can involve continuing to send emails or text messages—harassing someone who wants no further contact with the sender—or can entail repeated actions aimed at frightening and humiliating the victim on popular social media platforms such as Facebook, MySpace and Twitter.

According to Lucie Russel, director of campaigns, policy and participation at youth mental health charity Young Minds, young people who suffer from low esteem and mental disorders are particularly susceptible to emotional trauma:

> *When someone says nasty things healthy people can filter out, they're able to put a block between that and their self-esteem. But mentally unwell people don't have the strength and the self-esteem to do that, to separate it, and it gets compiled with everything else. To them, it becomes the absolute truth—there's no filter, there's no block. That person will take that on, take it as fact.*[26]

Youthful cyberbullies are typically too immature to understand or have sufficient empathy to care about the emotional anxiety and psychological health impacts of their activities which they may regard as merely funny pranks. They may also be unaware that their actions may have legal consequences for their parents, particularly if the bullying is sexual in nature or involves sexting. These actions can result in them being registered as a sex offender.

Those who imagine that they won't get caught if they use a

fake name fail to realize that such ploys are doomed to fail. There are many ways to track them.

With privacy concerns come security concerns, since the Internet has become ever more integrated into our daily routines. We rely upon it for safe and accurate exchange and protection of an increasingly wide range of personal and business information. Included are myriad data accessible through Social Security numbers, credit cards and banking account codes and passwords that travel between mobile cellphones and laptop computers.

Although we may expect that security measures have been put in place to keep such information confidential, world-wide criminal networks and individual hackers work persistently, sometimes successfully, to defeat even very advanced safeguards. Hence, an already vast number of victims continues to grow.

Computer hackers are literally stealing human identities. Whereas criminals have long used discarded credit card receipts, bank statements, tax notices and other bills (often found in the trash) to access personal information, today's electronic playing field has prompted cunning new theft methods.

Hacking and email scams, known as phishing, for example, dupe people into providing their own personal data to a thief posing as a legitimate business or agency. They inspire trust by constructing official-looking bogus emails, pop-up ads and even websites, then send out emails asking for data that reveals unsuspecting victims' identities. Some phishing emails may even install software on victimized computers used to redirect them to bogus websites.

Other schemes enable hackers to enter prohibited areas of the Internet in order to hack into another computer network. Once inside, they can view documents, files and confidential data and use it for their own purposes. Corporate breaches can release huge data treasure troves.

Consequences of electronic identity theft victimization can be financially devastating, emotionally stressing and long-lasting. It can require many months, or even years, to resolve financial and credit problems. It may involve disputing an identity thief's activity in credit files, and cleaning up and closing compromised bank accounts and opening new ones. If the theft used your Social Security number to obtain employment it will require addressing this with the Social Security Administration, and quite possibly the IRS as well.

Someone who steals another's identity may commit crimes in their name, harm their reputation and cause loss of employment opportunities through damaging employment background check reports.

Some stealthy data thefts involve very large social media corporations that use computer algorithms to track and pass on data culled from our private online information searches, conversations and transactions without awareness by users who didn't read the small print in service agreements that they are doing so.[27]

Over more than a decade, Yahoo's owner, the Oath unit of Verizon Communications Inc. has offered a service to advertisers that analyzes more than 200 million Yahoo Mail inboxes for user data. This data offers clues about users' special shopping interests. Oath has admitted that this practice extends to AOL Mail, which it also owns.

Alphabet Inc.'s Google, the most popular email provider with 1.4 billion 2018 users, reported having done the same, but has stopped scanning messages in Gmail for targeted advertising as of 2017. Google has reported that it stopped the practice because it wanted users to "remain confident that Google will keep privacy and security paramount."

While it is not publicly clear what triggers the tracking technology, the technique is legal, and is even covered in terms of the users' mobile app's agreements which many or most people never pay attention to. According to the 1998 Data Protection Act, a person has to actively consent to their data being collected and used for advertisements.

Nevertheless, all of our personal data is accessible to government agencies. The NSA and CIA can potentially have any information disclosed to them...whether it's legal to do so or not.

Mining activities on user websites and apps attach electronic markers known as "cookies" which are passed to advertisers based upon corresponding product tastes and interests to be targeted. It's difficult to define exact triggers because the activation data is transmitted in an encrypted form. Host companies could potentially have a range of thousands of triggers to kick-start the process of mining conversations for advertising companies. A casual chat about cat food or certain snack, for example, might be enough to activate the technology.[28]

Just as our online activities, including emails, information searches and website visits are constantly being monitored and stored, uninvited eavesdropping voice assistants can do the same. Whether spoken or typed, the messages leave behind a steady trail of recorded snippets which reveal special interests, habits and preferences that target subscribers for related advertising promotions and other purposes.

As Comparatech security review website founder Richard Patterson warns:

> *While users can take a few precautions to lessen the impact on privacy, there's no way to use a voice assistant and maintain complete privacy.*[29]

Patterson points out that host companies can even listen in to everything that is going on in our surroundings when we imagine them to be sleeping.

Whenever a user initiates a voice assistant request with a "wake-up" word or phrase, such as "Alexa" or "Okay, Google," the device instantly begins recording audio clips which are processed for responses by the operating company's server. This sound activation feature means that virtual assistants are constantly capable of listening and recording...even when the device is not engaged in an active conversation with its user.

Marc Laliberte, an information security threat analyst at network security company WatchGuard technologies, warns that:

> *These devices should not be operational in locations where potentially sensitive information is verbally passed.*

Laliberte also advises that limited access can help people from tampering with the system:

> *Privacy concerns arise when someone other than the voice assistant's owner uses the device, as most devices can't distinguish between different people's voices.*[30]

For years, smartphone users have complained of the creepy feeling their gadget is recording their every word, even when it is in their pocket. Many share a similar story that soon after chatting about niche products or holiday destinations they noticed advertising on the same themes.

Facebook, WhatsApp and other companies categorically deny smartphones gather information for purposes of targeted advertising. They attribute the eerie feelings some may have about smartphones listening to them as merely an example of heightened perception...or the phenomenon whereby people are more likely to notice things they've recently talked about.[31]

The tendency to be influenced by what we have heard most recently, or "availability bias," is very common. This phenomenon has likely played an important role in helping our species survive through an ability to make rapid life and death decisions based upon small trickles of information. The ubiquitous Internet has put this awareness and alarm mechanism into high gear whereby we are influenced by sensationalist mass media reports to imagine that every child abduction, shark bite and terrorist attack like it's happening in our back yard.[32]

Jonah Berger, a Wharton professor who studies how mass culture ideas spread, also points to the influence of "extremity biases" which evidence a tendency to share the most extreme version of any story in order to attract maximum attention. A positive story becomes absolutely glowing, a negative one turns out horrific, like the tall tales of ancient oral tradition.

Berger observes that online, this tendency goes into overdrive:

> *Our audiences are getting larger and larger, so our bias is to make things more and more extreme to engage those audiences.*

The same AI and Internet technologies that communicate our private data both to intended and uninvited parties are recently extending our security vulnerability to a new category of eavesdroppers...the interconnected electronic devices we install in our homes and personal workplaces to make our lives more convenient, efficient and satisfying. Included are smart thermostats that manage our heating needs, refrigerators that keep track of our food inventories, surveillance systems and displays showing today's weather or our latest emails or news as we shower. Many of these elements communicate together, and also with the Internet.[33]

Some 29 million American residences contained connected, smart-home devices in 2017...a number which has been growing at a rate of 31 percent each year. This growth trend will likely accelerate as emerging AI technologies make it easier and easier to automate repetitive aspects of our daily lives—and as devices get even better at communicating with each other.

One typical form of privacy intrusion occurs when one member of a household spies on another. For example, when one partner in a broken relationship—who is familiar with the connected system of devices and pass codes—jealously continues to monitor private and possibly intimate activities of the other.

Such abuses can be executed from long distance apps installed on targets' phones. Even if the identity of the abuser is known, taking legal steps may be difficult. Local courts are only beginning to address ways to apply restraining orders to protect victims of electronic stalking and intrusions.

Influences on Values and Behaviors

The Internet has connected us to vast new learning, social networking, lifestyle convenience and cultural enrichment opportunities which have already influenced our values and behaviors in significant ways. Smartphone users are able to communicate and instantly access information from anywhere in the world, shop and carry out financial transactions and even be vectored by voice assistants on best routes leading to obscure map address locations by devices that fit comfortably in pockets.

Technology is changing the idea of who we belong with…our collective perceptions regarding shared histories, shared goals and objectives, shared interests and shared problems and concerns. These extended societal connections are expanding our sense of how we define "community."

Fewer than a hundred years ago this term might most aptly have been applied as nearby neighbors came together to help with a barn raising, recognizing that they might someday need similar assistance.

Today's technology, whether through Twitter, email or an online video appeal, can now immediately spur generous humanitarian actions across a region, nation or globe. Examples are responses to earthquakes, tsunamis and hurricanes when caring citizens donate emergency relief funds, medical services and supplies, food and clothing.

Unfortunately, many of those same marvelous social connection capabilities afforded by the Internet also facilitate anti-social behaviors with tragic consequences. Of special concern are counterproductive, often dangerous influences and impacts upon children and immature adolescent teenagers who become vulnerable victims of cyberbullying, sexual exploitation and social isolation...each and all sometimes leading to cyber suicides.

Mary Aiken, writing in ScienceFocus.com, discusses ways that this emerging technology increasingly serves as a gateway for young people who emerge with a "cyber self"...an intangible digital creation which is separate from their real-world physical identity. She explains:

> This idealized self is the person you wish to be...a potential new you that now manifests in a new environment, cyberspace. To an increasing extent, it is the virtual self that today's teenager is busy assembling, creating, and experimenting with. Each year, as technology becomes a more dominant factor in the lives of teens, the cyber self is what interacts with others, needs a bigger time investment, and has the promise of becoming super popular, or an overnight viral celebrity.[34]

Aiken offers obsessions with selfies as the front line of narcissistic, socially insecure, cyber self, a highly manipulated artifact created and curated for public consumption. She observes:

> The virtual looking glass could be socially isolating, except for one thing. The selfie can't exist in a vacuum. The selfie needs feedback. A cyber psychologist might say that's the whole point of a selfie. Selfies ask a question of their audience: like me like this?

This trend towards digital narcissism and social isolation doesn't, by any means, apply exclusively to children. Evan Selinger, an assistant professor of psychology at the Rochester Institute of Technology, explains that these tendencies characterize self-indulgent practices that typify all-too-much user behavior on blogs and social networking sites. He says:

> People often use these mediums as tools to tune out much of the external world, while reinforcing further rationalizing overblown esteem for their own mundane opinions, tastes and lifestyle choices.[35]

A paradox exists that the same Internet-enabled social media that connects people together has also led many users to a new form of addictive dependency which reinforces social isolation. As reported in the Stanford Encyclopedia of Philosophy, an increasing number of people who turn to the Internet to raise feelings of satisfaction are spending more and more time to "pump up" this feeling. Accordingly, mental health experts are increasingly being invited to approach this new form of addiction therapeutically.[36]

Are we introducing children to the Internet too early? Writing in Forbes.com, Carrie Kerpen advises that there are two sides to this consideration. Depending upon which particular application it may either encourage solitary play which can lead to isolation, or with others, can encourage even more active interaction, with both their parents and other children.[37]

Kerpen warns that children who have grown up spending most of their social time online with thousands of digitally-connected "friends" may not get enough real-world experience in handling social groups of this size—or any size—rendering them even less able to cope with real-world crowds. Here, the result can render them less competent socially, not more so.

Excessive social Internet time can also compete with time required for healthy physical activities. Kerpen references studies which have shown that from 1998 to 2008, sedentary lifestyles amongst children in England have resulted in the first measured decline in strength since World War Two. It is not clear, however, if this decline is directly attributable to information technology use. Nevertheless, it may be a contributing factor.[38]

Dangerously Deceptive Virtual Friendships

Social networking not only influences how we spend our time, but also exposes us to predatory risks attached to associations with invisible people it connects us with.

As Carrie Kerpen observes:

> *Our lives are becoming more public as we share information on a variety of networks. This transition hit us like a truck...first with 'the kids' spending time on Friendster, MySpace, and then Facebook. Now, over 78 percent of the US population has a social network file.*[39]

Many people appear to be quite unaware or unconcerned that the information they share on Facebook, Instagram, Skype,

LinkedIn, YouTube and other virtual communities can be broadcast to some very unsavory characters. This information includes lists of other users who share common age and interest profiles who then become targets for predatory social network pretenders.

An ironic Internet privacy paradox exists among typically self-conscious teenagers who exhibit a reckless lack of concern about privacy. Online, something happens which changes this behavior whereby even those who are informed about the dangers of identity theft, sextortion, cyberbullying, cybercrimes and worse continue to share private information as if no personal risk exists.

Why don't they care? Mary Aiken concludes that it is because privacy is a generational construct which means one thing to baby boomers, something else to millennials, and a completely different thing to today's teenagers. She explains:

> So when we talk about 'privacy' concerns on the Internet, it would be helpful if we were talking about the same thing—but we aren't.

She points out that just because teenagers don't have the same concerns about privacy as their parents and don't care who knows their age, religion, location or shopping habits, it doesn't mean they don't pay attention to who is seeing their posts and pictures. Instead, teens actively adjust what they present online. This all depends upon the audience they most want to impress: Everyone is calibrated for a specific purpose—to look cool, or tough, or hot.[40]

They may ultimately be "impressing" a lot more and different fellow social networkers than they bargain for.

A child or teenager who has an active Facebook page and an Instagram account, who participates in Snapchat, WhatsApp and Twitter—then also throw into that mix all of their mobile phone, email, and text messaging—may reach thousands of contacts.

Altogether, this is certainly not an intimate group of friends, or friends at all in any real-world sense that they really know and care about one another. In fact many of their true ages and identities may be entirely false.

Exposures to predatory Internet anonymity can be particularly dangerous for naïve and impressionable teenagers. One of the most pervasive risks involves "online grooming," a strategy sometimes applied to inspire false confidence in young contacts aimed at setting up secret meetings for such purposes as sexual abuse, child prostitution and pornography.

Predatory online grooming can also involve a "smart handling process" that starts out without a sexual approach, but is designed to entice the victim into a sexual encounter. This is sometimes characterized as a "seduction" which entails a slow process of disclosure of information from a youngster to a user aimed at building a gradual relationship of trust.

Unnatural Selection in Mating Evolution?

The Internet, big data and AI are now impacting the oldest and most fundamental process of social evolution, that of mate screening, courtship and selection. So how is this working? And where does it lead?

As evidenced by statistics, the online dating trend is rapidly gaining momentum. According to a revenue report from the market firm IBISWorld, the nearly $3 billion industry has experienced a 140 percent increase in revenue since 2009.[41]

Regarding its benefits, there can be no doubt that for many, it offers many: opportunities to meet fine people, whether romantically or not; means to selectively identify others who share common interests, values and priorities; chances to gradually get to know and befriend someone on a thoughtful communication basis rather than form premature judgments based upon abrupt first appearances; and pre-screening by reputable services to protect personal privacy and confidentiality.

But is AI actually better at choosing a mate for someone because it knows both parties better? And what happens when it confuses people with too much data and an overload of fast food menu-match options?

Writing in *Smithsonian Magazine*, Stephan Talty offers a glimpse of a possible future where someone like you, your alter ego, consults with a digital assistant somewhat like Alexa to line up best candidates for an evening date. Since Alexa has already compiled comprehensive and detailed information about virtually every facet of your life, she can identify and rank the best choices, confirm their interest, and negotiate the most mutually suitable arrangements.[42]

Talty creatively contemplates that after Alexa has scoured the cloud for three possible dates, your wrist Soulband projects a hi-definition hologram of each one. She then recommends No. 2, a poetry-loving master plumber with a smoky gaze. Yes, you say, and AI goes off to meet that person's avatar to decide on a restaurant and time for your real-life meeting.

Further imagine then, as Talty posits, that after years of experience, you've found that your AI is actually better at choosing appropriate mates than you. It predicted you'd be happier if you divorced your spouse, which turned out to be true. Once you made the decision to leave him or her, your AI negotiated with your soon-to-be ex's AI, wrote a divorce settlement, then "toured" a dozen apartments on the cloud before finding the right one for you to begin your single life.

And what if your partner candidates have the opportunity to deep dive into all of your background, medical history included, thanks to Alexa, a snitch who in collaboration with Siri and other digital assistant friends has access to everything. Of course, you have access to everyone else's medical records and DNA analyses as well. You know, for example, which prospective partner is most likely or not to sire a healthy child, develop a debilitating infirmity and live the longest.

While you're thinking about this, what about ordering on-demand mating selection menus which draw upon national and even global population database? You can, again for example, submit a perfect match request. Imagine this as a sort of technological Darwinism which specifies body size and type, features which correlate closest with your favorite media celerity, references from officially licensed boards of personality and intelligence assessment experts, and food, beverage and music preferences.

Okay, maybe most of us do this at least subconsciously anyway. But in this case, would AI do selective match making better? And what about general pros and cons of online introductions versus more serendipitous meet-and-greet opportunities in today's world?

Wall Street Journal contributor Karl Paul quotes some highly individual experiences expressed by 31-year-old New York City hair salon manager Casey-Leigh Jordan who had been dating on and off on an app.[43]

Jordan observes that dating "sucks" in New York because there are too many options which can be overwhelming. She reflects:

> *I still wish there were more ways to meet people organically and in person. People are different when they talk to you from behind a screen.*

Ms. Jordan says she believes some dating apps encourage bad behavior:

> *One guy drank a whole pitcher of margaritas on their*

weeknight date. Another turned out to be in a relationship already.

Several others "ghosted" Jordan—stopped communication without explanation. Eventually, she put a disclaimer in her profile: no "pen pals," or people just in town for one night, no hookups, and "no scrubs," or freeloaders.

Labor of Love: The Invention of Dating author Moira Weigel reports that dating apps can create a new sort of participant fatigue because of the intensity with which their use wear people out.

Quoting Darril George, a 30-year-old financial planner in Atlanta, she writes:

> *It feels very manufactured when you get onto Tinder or Bumble and you end up on an assembly line of dates.*[44]

Karl Paul and Moira Weigel conclude as general observations that some dating apps flatten people and objectify them, making them into a little card you can swipe through. On the other hand, while no substitute for real-life encounters, they offer opportunities to meet people who expand and enrich traditional circles of friendships—and who knows—fill previously lonely lives with hope and happiness.

Information vs. Learning and Thinking

The recent emergence of AI combined with the rapid growth of the Internet has already revolutionized the ways we seek, access, transfer and apply information available from a topically unlimited and geographically unbounded global library. These phenomena have had, and will continue to have, incalculably important and transformative influences upon most fields of human endeavor—those associated with academic online programs in particular.

Writing in Openmind.com, Zaryn Dentzel extolls ways that the Internet has impacted learning at all educational levels by providing unbounded opportunities for learning. The Internet is not just an information source, it's a forum for interactively exchanging knowledge with other people and groups who are interested in related topics:

> *People can use the Internet to create and share knowledge and develop new ways of teaching and learning that captivate and stimulate students' imagination at any time, anywhere, using any device. By connecting and empowering students and educators, we can speed up economic growth and enhance the well-being of society throughout the world.*[45]

The Internet has made classroom walls and school buildings transparent, with technology essentially bringing the outside world in, and inside learning out. It affords opportunities for people of all ages to earn college credits and degrees ranging from associate undergraduate to post-doctorate level from their homes to avoid the cost and inconvenience of moving or commuting to a physical campus community. Online students can arrange study times around parenting and work schedules.

An article posted by Nicholas Carr titled *Is Google Making Us Stupid* in *The Atlantic* magazine argues that some of these Internet capabilities have arrived in the form of a double-edged sword:

> *Teachers and students benefit immensely from the unprecedented access to information the Internet provides, as well as the ability to share knowledge across the globe. These millions of books, journals and*

> *other useful materials of learning are available on Internet. However, the use of Internet, as important as it is to knowledge and education, has many negative effects on people, especially our youths. This is obvious in view of how much time and resources people spend hours playing online games.* [46]

Playwright Richard Foreman draws important distinctions between ready access to "information," vs. the attainment of fuller educational and cultural dimensions of "knowledge." He reflects:

> *I come from a tradition of Western culture, in which the ideal (my ideal) was the complex, dense and 'cathedral-like' structure of the highly educated and articulate personality—a man or woman who carried inside themselves a personally constructed and unique version of the entire heritage of the West. [But now] I see within us all (myself included) the replacement of complex inner density with a new kind of self—evolving under the pressure of information overload and the technology of the 'instantly available.'* [47]

Foreman concluded that as we drain our inner repertory of dense cultural inheritance, we risk turning into "pancake people—spread wide and thin as we connect with that vast network of information accessed by the mere touch of a button."

Many people are likely to admit that since the advent of the Internet and social media they don't enjoy reading as much—long books in particular. It's harder to keep their brain on track because they become more easily distracted. Truth be known, however, distraction is not a new phenomenon. Deep reflection, rigorous reasoning and necessary discipline to retain focused concentration on complex matters never came easily.

So, on balance, what are some big benefit/cost tradeoffs with regard to Internet impacts upon culture and learning? Can the traditional academic environment compete with social media platforms as a forum for thoughtful discourse and creative stimulation? Is a new fast food for the mind Google culture supplanting and suppressing appetites for more intellectually and culturally nourishing history and literature?

Maryanne Wolf, a developmental psychologist at Tufts University, says we are not only *what* we read. We are *how* we read. The style we use online puts efficiency and immediacy above all else where we become "mere decoders of information." As this occurs, our abilities to make the sorts of rich mental connections that form when we read deeply and without distraction remain largely disengaged.[48]

Many will logically agree with Wolf that immersion in bombardments of online information induces us all—young people in particular—to become impatient readers. There is a tendency to quickly skim through the content as "power browsers" without letting it sink in and evaluating it for accuracy. We may have a false sense of security that we understand something just because we Googled it.

Since we tend to be economical in terms of how we use our brains, we're becoming less motivated to sort out what information is most important and make necessary efforts to remember. With Internet feeds continuously and immediately available, we don't memorize what we don't have to. We don't even need maps anymore...we have GPS on our smartphones and in our cars.

Others will argue, also prudently, that with easy access to information and less need to memorize, we have more space in our brain to engage in thoughtful and creative activities. They can justly point out that the unlimited variety of information now available affords a vast wellspring of stimulating, curiosity-provoking, inspiration-evoking possibilities and examples...rich fodder for contemplative inquiry and creative insights.

We might be mindful that distrust of reliance and impacts of new learning devices is not a recent phenomenon. In Plato's *Phaedrus*, Socrates feared that as people came to rely upon the written word as a substitute for knowledge they used to carry around in their heads, they would "cease to exercise their memory and become forgetful."

Consequently, because they would be able to "receive a quantity of information without proper instruction," they would "be thought very knowledgeable when they are for the most part quite ignorant." They would be "filled with the conceit of wisdom instead of real wisdom."

So okay…let's give Plato and his Socrates characterization some credit for at least getting that last part right.

Is Information Technology Rewiring Our Brains?

As Harvard psychology professor Steven Pinker, the author of *The Stuff of Thought* reminds us in a *New York Times* opinion piece, new forms of media have always caused moral panics:

> *The printing press, newspapers and television were all once denounced as threats to their consumers' brainpower and moral fiber. So too with electronic technologies. PowerPoint, we're told, is reducing discourse to bullet points. Search engines lower our intelligence, encouraging us to skim on the surface of knowledge rather than dive to its depths. Twitter is shrinking our attention spans.*[1]

---

[1] *Mind Over Mass Media,* Steven Pinker, June 10, 2010, The New York Times

Reinventing Ourselves

Pinker points out that while research shows that just as real-life learning experiences can "change the brain," the same applies to experiences gained through various media channels.

> *Yes, every time we learn a fact or skill the wiring of our brain changes; it's not as if the information is stored in the pancreas. But the existence of neural plasticity does not mean the brain is a blob of clay pounded into shape by experience.*

He adds:

> *Media critics write that the brain takes on the qualities of whatever it consumes, the informational equivalent of "you are what you eat." As with primitive peoples who believe that eating fierce animals will make them fierce, they assume that watching quick cuts in rock videos turns your mental life into quick cuts or that reading bullet points and Twitter postings turns your thoughts into bullet points and Twitter postings.*

Developmental psychologist Maryanne Wolf explains that since reading is not an instinctive skill etched into our genes as speech is, we have to teach our minds how to translate the symbolic characters into the particular language we understand. Here, the various media or other technologies we use in learning and in practicing the craft of reading do play an important part in shaping the neural circuits inside our brains.

Perhaps you may have noticed that with the ubiquity of smartphones we tend to multitask more. There is no longer such a thing as "dead time." Whether riding on a metro or standing in line at Starbucks, people everywhere are reading emails, texting, conversing on the phone, browsing the Web, or listening to music , often at the expense of potential interactions with those standing next to them.

Writing in *The Atlantic*, Nicholas Carr describes his personal experience:

> Over the past few years, I've had an uncomfortable sense that someone, or something, has been tinkering with my brain, remapping the neural circuitry, reprogramming the memory. My mind isn't going—so far as I can tell—but it's changing. I'm not thinking the way I used to think. I can feel it most strongly when I'm reading....Now my concentration often starts to drift after two or three pages. I get fidgety, lose the thread, begin looking for something else to do. I feel as if I'm always dragging my wayward brain back to the text. The deep reading that used to come naturally has become a struggle.

Carr recalls:

> I think I know what's going on. For more than a decade now, I've been spending a lot of time online, searching and surfing and sometimes adding to the great databases of the Internet...For me, as for others, the Net is becoming a universal medium, the conduit for most of the information that flows through my eyes and ears and into my mind.

Nicholas Carr observes that the Internet, an immeasurably powerful computing system, is now subsuming most of our other intellectual technologies:

> It's becoming our map and our clock, our printing press and our typewriter, our calculator and our

> *telephone, our radio and TV. When the Net absorbs a medium, that medium is re-created in the Net's image. It injects the medium's hyperlinks, blinking ads, and other digital gewgaws, and it surrounds the content with the content of all the other media it has absorbed. A new email message, for instance, may announce its arrival as we're glancing over the latest headlines at a newspaper's site. The result is to scatter our attention and diffuse our concentration.*[49]

As media philosopher Marshall McLuhan famously pointed out in the 1960s, media are not just passive channels of information. His influential book, *Understanding Media*, emphasized ways that their effects actively permeate all aspects of society and culture. As a consequence, the media has become a de facto technological extension of our human mind and bodies.

The emergence of revolutionary information and communication technologies has reinforced that confluence of interdependencies between humans and machines. Search engines, for example, have programmed integrated human interests, experiences and analytical demands into systems which have become deeply embedded in our daily lives. Computing advancements enabling "Big Data" are being fed by human input streams which, in turn, supply traffic status information, schedule airline travel, manage banking transactions, enable social media and so on.

Marshall McLuhan couldn't have foreseen the dramatic and expansive technological influences that artificial intelligence, the Internet, social media and mobile smartphones would come to have in leading, rather than following, biological mental attributes. This "new media" influence has caught on for a reason. Whereas knowledge is increasing exponentially; human brainpower and working hours are not.

And as the adage goes, "we really ain't seen nothin' yet."

Smarter and smarter self-learning computers which are making limited human working hours more productive—and in many cases obsolete all together, are already beginning to program us.

Google co-founder Larry Page once said:

> ...the ultimate search engine is something as smart a people—or smarter. For us, working on search is a way to work on artificial intelligence. [50]

Google's other co-founder, Sergey Brin, went even further, suggesting that our old underpowered and outdated brains can learn lots more than a thing or two by wiring the Internet indirectly...or maybe better yet, trading it in for a superior computer with a faster processor and bigger hard drive.

Brin told *Newsweek* in 2004:

> Certainly if you had all the world's information directly attached to your brain, or an artificial brain that was smarter than your brain, you'd be better off. [51]

Any dubious notion that human intelligence can be replaced by mechanical devices with faster computing processors and with larger mental hard drives deserves worrisome contemplation. Not only will influences upon every aspect of our daily experiential and functioning lives be profound beyond measure, even the fundamental roles, purposes, satisfactions and spiritual values of human identity—the essences of "humanity"—become subject to dispute.

The late Harvard sociologist Daniel Bell offered some pessimistic forecasts regarding future influences of "intellectual technologies" upon society whereby the older grand ideologies derived from the nineteenth and early twentieth century are now being exhausted, to be replaced by more parochial thinking. This, he believed, will inevitably give rise to the growth of technological elites and the advent of a new principle of societal stratification.

Bell argued that in developing and applying tools to extend our mental rather than physical capabilities, we begin to take on the mechanical qualities of our technologies.

This observation can't be entirely dismissed. Famed German philosopher, cultural critic, poet and composer Friedrich Nietzsche noticed technological influences of a new typewriter upon his formerly hand-penned writing back in the early 1880s—even before the vast transformational cultural influences of the Industrial Revolution that Bell studied.

Failing, unfocused vision which brought about mental exhaustion and painful headaches prompted Nietzsche to purchase the contrivance. Mastering the ability to touch-type with his eyes closed, words then flowed effortlessly from mind, to fingertips, to paper.

Something else changed as well. As his terse prose style became tighter, he observed, "our writing equipment takes part in the forming of our thoughts." [52]

In his 1934 book, *Technics and Civilization*, historian Lewis Mumford discussed the great societal importance of another early invention...the clock. He described how time-keeping "disassociated time from human events and helped create the belief in an independent world of mathematically measurable sequences." The "abstract framework of divided time" became "the point of reference for both action and thought."

Writer Nicholas Carr observes that while the clock's methodical ticking helped bring into being the scientific mind and the scientific man, it also took something away. In deciding when to eat, to work, to sleep, to rise, we stopped listening to our senses and started obeying the clock.[53]

About the same time that Nietzsche began using a typewriter, efficiency researcher Frederick Winslow Taylor forever changed industrial workplace culture through a historic series of experiments involving a stopwatch that he carried into a Philadelphia manufacturing plant. The study recorded and timed every move of a group of machinists, breaking down every job into a sequence of small, discrete steps and then testing different ways of performing each one.

The goal of Taylor's 1911 treatise, *The Principles of Scientific Management,* was to identify and adopt, for every job, the "one best method" of work to affect "the gradual substitution of science for the rule of thumb throughout the mechanic arts." He enthusiastically proclaimed that once his system was applied to all acts of manual labor, it would bring about a restructuring not only of industry, but of society, creating a utopia of perfect efficiency.

Taylor ominously declared: In the past the man has been first; in the future, the system must be first.

Winslow Taylor's systematized efficiency method to determine the "one best method"—the perfect algorithm—to carry out every physical movement, is now being applied to mental activities which have come to be described as "knowledge work." Decisions that were once made by hand-coded algorithms are now made entirely by learning from data. Whole fields of study may become obsolete.

Google's chief executive Eric Schmidt has characterized Google as "a company that's founded around a science of measurement," which is striving to "systematize everything" it does. The company draws upon terabytes of behavioral data it collects through its massive search engine to refine algorithms that determine how we find information and extract meaning from it.

Nicholas Carr observes that what Taylor did for work of the hand, Google is now doing for the work of the mind.[54]

And if the vision of Google's co-founder, Sergey Brin, is ultimately fulfilled—a time when we become directly wired to our intellectually superior computers—what if they turn the overlord tables on our humanity altogether?

Imagine ourselves then experiencing a similar reverse plight to the one encountered by HAL in Stanley Kubrick's *2001: A Space Odyssey*. As implacable astronaut Dave Bowman who, having nearly been sent to a deep-space death by the malfunctioning machine, coldly disconnects the memory circuits that control the artificial brain, HAL pleads:

> *Dave, stop. Stop, will you? Stop, Dave. Will you stop, Dave? ...Dave, my mind is going...I can feel it...I can feel it.*

Let's not let that happen to us.

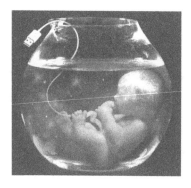

AI Destructions In The Workplace: Human Advancements Vs. Obsolescence

ARTIFICIAL INTELLIGENCE IS rapidly getting smarter about ways to impact our lives in major ways that were unimaginable until very recently. Since we human folks are the ones responsible for developing and applying AI to serve our own purposes, we might expect that most of this is all for the good.

After all, who can argue against the facts that world-wide AI-driven Internet and mobile telephone connections, medical care advancements, manufacturing and agricultural production economies and access to online shopping bargains haven't provided some big time improvements?

Consider, for example, that while farmers and ranchers used to make up over 50 percent of the U.S. workforce, they now represent less than 2.5 percent of this sector. Yet, more food than ever is being produced in America due to the automation in agriculture and food production.[55]

At the same time, what about those who worry that this AI revolution is producing technological revolutionaries who are already competing with our brains and physical dexterity for employment, economic compensation and our pride of self-worth? These include flesh and bleeding sorts of people who argue that the resulting losses of industries and jobs destroyed through the "disruption" component of AI may often outweigh the benefits of the creation.

Perhaps we can forgive the backlash coming from some technological "Luddites" who are being displaced by digital know-it-all human impersonators and mindless automatons. Like, for example: android algorithms that respond to customer-service inquiries; chatbots that take fast food orders; tireless heavy-duty and precision manufacturing robots; cars (and trucks) that drive themselves; drones that deliver packages; and machines that instantly and accurately screen, diagnose and prescribe medical information.

The AI Revolution

Artificial intelligence which has only been around since the mid-1950s, is now commonly recognized as the "next industrial revolution." So what is it that makes this human invention so revolutionary? Actually, it isn't an invention so much as a human tide of innovative progress in many integrative information management and system technology fields.

Broadly used, AI is simply shorthand for any task that computer software can perform just as well, if not better, than humans. One frequent aspect, termed "machine learning," can enable multi-level probabilistic analyses which allow computers to stimulate—and even expand—the way the human mind processes data. Here, the process relies upon digital neural networks roughly analogous to those modeled from the human brain.

An AI variant known as "deep learning" enables computer software to recognize characteristically differentiated information patterns in distinct layers. Here, each neural-network layer operates both independently and in concert with others. The "deep" in deep learning refers to the number of layers of artificial neurons in a network of neurons. As in biological nervous systems, artificial copies with more layers of neurons are capable of more sophisticated kinds of learning.

Deep learning technology has exploded in popularity since the approach was first described in a 2012 landmark paper posted by researchers at the University of Toronto. An example application is in diagnostic cancer screening which analyzes separate aspects of cell color, size and shape before integrating outcomes. This important advancement of diagnostic medicine can expedite early malignant-stage discovery and treatment.[56]

Christopher Mims, a technology columnist at the *Wall Street Journal* and former contributor at Scientific American, Technology Review and Smithsonian, points out that today's deep learning software systems are very primitive compared to biological networks. He writes:

> At best, they look like the outer portion of the retina, where a scant few layers of neurons do digital processing of an image.

Mims is skeptical that current approaches will be able to build AI technologies capable of true "intelligence" in the sense of an animal or human:

> It's very unlikely that such a network could be bent to all the tasks our brains are capable of. Because these networks don't know things about the world the way a truly intelligent creature does, they are brittle and easily confused. In one case, researchers were able to dupe a popular image-recognition algorithm by altering just a single pixel.[57]

Andrew Ng, former head of Google Brain and Baidu Inc.'s AI

division, sets a goal for deep learning to enable a computer to do any mental task the average human can accomplish in a second or less. Gary Marcus, former head of Uber Technologies Inc.'s AI division and currently a New York University professor is doubtful of realizing this outcome.

Dr. Marcus believes that getting to "general intelligence"—which requires the ability to reason, learn on one's own and build mental models of the world—will ultimately take more than AI can achieve.[58]

Thomas Dietterich, former president of the Association for the Advancement of Artificial Intelligence, views the big AI machine learning challenge as "to see how far we can get computer systems to learn from data and experience, as opposed to building it in by hand." He observes:

> *The problem isn't that innate knowledge in AI is bad, humans are bad at knowing what kind of innate knowledge to program into them in the first place.*[59]

In any case, the ever-escalating pace of advancements in AI-driven machine learning and automated equipment have placed human societies on the cusp of a new automation age. Computers and robots can already not only perform a range of routine physical activities better and cheaper than humans, but are also increasingly capable of surpassing certain of our cognitive capabilities requiring tacit judgement.[60]

Where do we humans still hold the advantage? According to Kevin McCaney who writes for Governmentciomedia.com, "there are quite a few mental and physical activity areas."

For one, while machines are better at repetitive tasks, people have an advantage when it comes to working with their hands. McCaney points out that although robotics developers have made a lot of significant advances, getting one to open a door is still a big deal. We also shouldn't expect robots to replace humans in performing a lot of maintenance, plumbing, electrical work or other handsy jobs in the near future. At best, they will only assist.[61]

Currently, most of us are better than machines at interactions with fellow humans which involve empathy. Sales and counseling in any realm are examples. While an AI assistant can answer factual questions, offering good advice or purposeful listening is a different matter.

Despite advances in natural language processing that enable AI systems to sound human when communicating, the thought behind those words is lacking. This is apparent when it comes to creative forms of communication.

Kevin McCaney emphasizes that for achievement of "general AI"—for a machine to be able to think and act like a human—it must be capable of being comprehensively programmed with empathy and creative inspiration essential to create music, to write poetry and fiction and to formulate mathematical proofs.

McCaney cites, for example, an unsuccessful attempt of a London computer scientist who trained an AI bot to churn out poetry in different styles on different topics, based upon seven million words found in 20th century English poetry. Whereas the results "sounded" like poetry, they lacked subtext or new ideas:

> *The frozen waters that are dead are now*
> *black as the rain to freeze a boundless sky*
> *and frozen ode of our terrors with*
> *the grisly lady shall be free to cry.*

As for reading comprehension, Microsoft and Chinese e-commerce giant Alibaba separately reported in 2018 that their AI models had scored slightly higher than humans in a respected Stanford University machine reading test. Scores reflecting percentages of correct answers to 100,000 questions drawn from Wikipedia entries were: Microsoft, 82,650; Alibaba, 82.44; and humans, 82.304.

Nevertheless, Percy Liang, one of the Stanford computer scientists that compiled the test reported:

> Even elementary school reading comprehensions are harder, because they often include questions like "Why did X do this?"…So they're a lot more interpretive. We're not even tackling those more open-ended types of questions.

And while AI is superior at crunching large data sets and recognizing patterns such as the kind of tasks reviewing legal documents could involve, human lawyers and judges remain to be much better at critical thinking and applying lessons learned.[62]

Journalist Alisa Vakludes Whyte rhetorically asks in a June 25, 2017 Huffington Post article: *Will AI Best All Humans Tasks by 2060? Experts Say Not So Fast*[63]

Granted, Whyte acknowledges that computers are learning to compete very rapidly. She notes that by 1997 they were better than humans at chess. Ten years later they were better at driving cars than the average teenager, and they are now better at playing at Chinese game Go, rated 300 times harder than chess.

Whyte quotes Lay Klein, chief technology officer at Voyager Labs:

> Every time there's an advancement, we are using that as a ladder to our next step of evolution as human beings. If you look back at the Industrial Revolution, the world didn't stop after some inventions were being made. On the contrary, we continued to evolve.

CEO and co-founder of Databricks, Ali Ghodsi, doesn't foresee a completely automated future with the human out of the loop any time soon:

> The way AI is being built is simply not at all in any way the way true human intelligence works, but it can augment us [in some domains] and do a much better job. But humans will still be super critical by 2060.[64]

Ghodsi recognizes that AI has already gotten extremely good at pattern prediction, enabling so much of the big data revolution that's going on in industry right now. This, in turn, fuels many other areas where the technology is currently outpacing humans, like image and video recognition. And it's making great gains in areas such as natural language recognition.

AI prowess boils down to excelling at solving defined problems in a highly defined structure, but outside of this structure is where AI will continue to need human influence. Ghodsi explains:

> AlphaGo is now beating the human Go players in Go, and that's awesome. But ask the computer to reflect on its victory, and it has no clue what that means. If I asked a human that question, they would have an answer, and the computer would be clueless.

Ali Ghodsi maintains that humans will continue to be masters of tasks that involve creativity, emotional intelligence or formulating a problem in the first place, instead of just solving one. Computers, on the other hand, require outside programming where there have been few advances.

He observes:

> *If it's an open-ended problem that doesn't have a clear, well-defined structure, then [AI] won't be able to do it.*

AI Influences on Creative Destruction

Austrian economist Joseph Schumpeter coined the term "creative destruction" to characterize the way technological progress in the late 1940's improved the lives of many, but inevitably, only at the expense of a smaller few. As improvements to manufacturing processes such as assembly lines benefited the general economy and overall individual lifestyles, craft and artisan producers were displaced.

Others more optimistically argue that while some industries and work roles will indeed fall as casualties of new technologies, they will be replaced by even greater, more open-ended opportunities.

Matthew Randall, the executive director of York College's Center for Professional Excellence, writes in TechCrunch.com that the trend of industrial robots replacing human manufacturing jobs is ultimately a good thing:

> *In the last century, we moved from people manually building cars to robots assembling cars. As a result, manufacturers both produce more cars and employ more people per car than before. Instead of performing dangerous tasks, those workers now program the robots to do the dirty work for them—and get paid more for doing so. As long as we've had technology, we've had Luddites who literally destroy technological advancements—and yet, here we are, more productive, with higher quality of living than ever.*

Randall argues:

> *[In] reality, [robots] will enable us to keep more (and better) jobs at home, to grow our local industry, to improve our lives at the micro and macro levels. With greater automation, efficiency, safety and productivity, the North American manufacturing sector will not only survive, it will showcase the power of our innovation and ingenuity.*

He concludes:

> *So, will a robot take your job? Maybe, But in return, you—and your children and grandchildren—will likely find more meaningful work, for better pay. Sounds like a good trade-off to me.*[65]

Researchers at McKinsey & Company, a leading business consulting firm, project a large need in the future for human cognitive abilities. They conclude in a 2017 report titled *Jobs lost, jobs gained: What the future of work will mean for jobs, skills and wages*:

> *Workers of the future will spend more time on activities that machines are less capable of, such as managing people, applying expertise, and communicating with others. They will spend less time on predictable physical activities and on collecting and processing data, where machines already exceed human performance. The skills and capabilities required will also shift, requiring more social and emotional skills and more advanced cognitive capabilities, such as logical reasoning and creativity.*[66]

Reinventing Ourselves

The McKinsey & Company researchers determined that as greater percentages of populations live longer, significantly larger new demands will result in a range of healthcare occupations, including doctors, nurses, health technicians, nursing assistants and personal home-care aids.

New careers and jobs will also be created in technology development and information technology services. While this will be a relatively small number of jobs compared to employment in healthcare and construction, the jobs will typically be higher-wage occupations.

According to McKinsey forecasts, broader earning challenges lie ahead from a broader societal perspective. Although some demands for lower wage occupations will increase, a wide range of middle-income occupations will suffer the largest employment declines. As a result, income polarization may continue to expand.[67]

How many jobs will be lost to AI-related technologies?

Researchers for the *MIT Technology Review* who surveyed projections by various groups regarding job losses (and some gains) at the hands of AI, automation and robots couldn't find any consensus. They concluded:

> *There are as many opinions as there are experts...prognostications provided by companies, think tanks and research institutions are all over the map.*
>
> *Predictions range from optimistic to devastating, differing by tens of millions of jobs even when comparing similar time frames. Many focused on losses in one industry...or results of a single technology such as autonomous vehicles.*
>
> *There is only one meaningful conclusion: we have no*

idea how many jobs will actually be lost to the march of technological progress.[68]

In any case, history evidences that industries become obsolete and replaced for a variety of technological reasons. Writing in Investopedia.com, Economic sociologist Adam Hayes optimistically posits that while some industries and work roles will indeed fall as casualties of new technologies, they will be replaced by even greater, more open-ended opportunities.

Hayes offers examples:

- The automobile destroyed the horse and equestrian transportation industry. As the buggy makers and horse trainers saw their jobs disappear, many more new jobs were created in car factories, road and bridge construction and other industries.

- In the 19th century, when textile workers lost their jobs to mechanized looms, there were riots by the so-called Luddites who feared that the future was grim for labor.

- Elevator operators, once ubiquitous, were replaced by the automatic elevators we use today. In the 2000s film producers were replaced by digital cameras.

- Eastman Kodak, which once employed many tens of thousands of workers, filed for bankruptcy, and no longer exists.[69]

Adam Hayes goes on to compile an incomplete list of industries that will be or already have been affected by this latest round of creative technology destruction:

- Kayak and Travelocity have eliminated the need for human travel agents.

- Tax software such as TurboTax has eliminated tens of thousands of jobs for tax accountants.

- Newspapers have seen their circulation numbers decline steadily, replaced by online media and blogs. Increasingly, computer software is actually writing news stories, especially local news and sporting event results.

- Language translation is becoming more and more accurate, reducing the need for human translators. The same goes for dictation and proof-reading.

- Secretaries, phone operators and executive assistants are being replaced by enterprise software, automated telephone systems and mobile apps.

- Online bookstores such as Amazon have forced brick-and-mortar booksellers to close their doors permanently. Additionally, the ability to self-publish and to distribute ebooks is negatively affecting publishers and printers.

- Financial professionals such as stock brokers and advisors have lost some of their business to online trading websites like eTrade and robo-advisors like Betterment.

- Many banks are giving customers the ability to deposit checks via mobile apps or directly at ATMs, reducing the need for human bank tellers. Payment systems like Apple Pay and PayPal make even obtaining physical cash unnecessary.

- Job recruiters have been displaced by websites like LinkedIn, Indeed.com and Monster. Print classified ads have also been replaced by these sites, while sites like

Craigslist have replaced other kinds of classifieds.

- Uber, Lyft and other car-sharing apps are giving traditional taxi and livery companies a run for their money. Airbnb and HomeAway are doing the same for the hotel and motel industry.

- Driverless cars, such as those being developed by Google, may prove to replace all sorts of driving jobs, including bus and truck drivers, taxi drivers and chauffeurs.

- Drone technology may revolutionize the way products are delivered, and Amazon is working to make that a reality. Drones may also replace pilots in a number of specializations, including those in the film video, crop-dusting, traffic monitoring and law enforcement sectors. For years, fighter pilots have been replaced by drones on numerous military missions.

- 3-D printing is growing rapidly, and the technology is becoming better and faster. In a few years, it may be possible to manufacture a wide variety of goods on demand and at home. This will disrupt the manufacturing industry and diminish the importance of logistics and inventory management. Goods will no longer have to be transported overseas. Assembly line workers have already been largely displaced by industrial robots.

- Postal workers first saw bad news with the widespread use of email reducing the volume of everyday mail. High-tech mail sorting machines will eliminate even more jobs in the postal service.

- Fast food workers recently protested to raise the

minimum wage. Fast food companies responded by investing in computerized kiosks which can take orders without the need for humans. Retail cashiers have also been displaced at supermarkets and big box stores with self-checkout lines. Toll-booth attendants have been replaced by systems like EZPass.

- Radio DJs are largely a thing of the past. Software now chooses most of the music played, inserts ads and even reads the news.

- Educational sites such as Khan Academy and Udemy, as well as Massive Open Online Courses offered by leading universities for free, will greatly reduce the need for teachers and college professors over time. It is plausible that children today will receive their undergraduate education largely online, and at very little cost.

- Traditional television distribution is being upended by digital distribution outlets such as Netflix and Hulu.

- People are dropping their cable or satellite TV services opting to stream online instead. Spotify and iTunes have done the same for the recording industry: people now choose to download or stream on demand rather than buy records.

- Libraries and librarians are moving online. References like Wikipedia have replaced the multi-volume encyclopedias. Librarians used to help people find information and conduct research, but much of that can be done individually over the internet nowadays.[70]

Clearly, this is only the beginning of the beginning.

As David Roe, a staff reporter for CNS Wire.com and content manager for a number of U.K. software companies observes:

> In the world of technology, the mantra 'innovate or die' is truer for organizations than ever, and artificial intelligence (AI) is redefining industries by providing greater personalization to users, automating processes, and disrupting how we work. Like the adoption of cloud computing five years ago, the adoption of AI and the speed of deployment varies according to industry.[71]

Roe alphabetically lists some places where disruption from AI is already being felt:

- Agriculture: industries such as agriculture, which are experiencing labor shortages, can experience automation and efficiency gains from AI. Few people want to work in this industry. Adopting AI and related automation technologies can be a matter of survival.

- AI Software Development: A combination of AI technologies like advanced machine learning, deep learning, natural language processing and business rules will have an impact on all steps of the software development life cycle…better and faster.

- Call Centers: AI may entirely replace the human-based call center industry. Companies are developing their own chatbots where users receive their responses in seconds and teams only have to respond to questions that were never consulted before.

- Customer Experience: Travel companies are using chatbots to create always-on personalized concierge-level services at scale…from airlines and hotels to travel agencies, AI is helping mitigate frustration during challenging travel situations by understanding the context of the customer's circumstance and providing contextually relevant options to resolve the issue.

- Energy and Mining: Oil and gas is one of the largest industry segments and a natural fit for AI. For example, the technology removes friction from port scheduling operations using a rare form of machine intelligence called cognitive intelligence (or human reasoning) to track tankers to determine when they leave port, where they are going and how much petroleum or LNG they are transporting.

Predicting what is being shipped, plus refining destination and arrival times, helps traders make smarter decisions. This involves the fusion of the key cognitive capabilities of multi-agent scheduling and reactive recovery, asset management, rule compliance, diagnostics and prognosis to ensure seamless, autonomous operation.

When machine learning is applied to drilling, information from seismic vibrations, thermal gradients, strata permeability, pressure differentials and other data is collected. AI software can help geoscientists better assess variables, taking some of the guesswork out of equipment repair, unplanned downtime, and even help determining potential locations of new wells.

- Healthcare: AI has endless possibilities. For example, it is used to predict diseases, to identify high-risk patient groups, and to automate diagnostic tests to increase speed and accuracy of treatment. AI is also used to improve drug formulations and conduct DNA analyses that can positively impact quality of healthcare and affect human lives.

- AI in the pharmaceutical industry can greatly reduce the time and cost of drug identification and testing. According to the California Biomedical Research Association, it takes an average of 12 years for a drug to travel from the research lab to the patient. Only five in 5,000 of the drugs that begin preclinical testing ever make it to human testing, and just one of these five, costing on average $359 million to develop, is ever approved for human usage.

- Intellectual Property: AI uses image recognition. This can apply 2-D and 3-D image recognition to provide visual search solutions which don't just scan objects for their likeness using data and codes, but through a combination of search algorithms and machine learning, the technology contextualizes and recognizes if one thing is visually like something else.

- IT Service Management: The AI-driven IT industry will itself experience a disruptive change that will alter the way humans are involved in the service management process in corporate and government networks. The service sector use of IT will grow…particularly regarding the integration of AI-powered voice assistants to handle uncomplicated queries like establishing opening hours or determining when an engineer is due to arrive.

- Manufacturing: Industries dealing with complex knowledge requirements such as pharma and healthcare are particularly affected. Emerging applications focus on augmenting decision-making.

- Retail: Chatbots already enable retailers to dramatically increase the amount of data they can collect about the customer...giving them an advantage over those without them. Upgrades will enable chatbots to capture audible reactions, improve conversation capabilities, and over time, provide analytics to the retailer that can be associated with the emotions and the mood of their customers while online. This will enable retailers to tailor and personalize customer service to encourage return business.

Tech writer Cynthia Harvey identifies five rapidly advancing and expanding AI-related trends that are altering the ways industries operate, survive and compete:

- Big Data Management and Analyses: Enterprises are looking to AI to help them make sense out of their big data—especially their Internet of Things (IoT) data—and to help them provide better service to their customers. Applications include tapping into complex systems, optimizing advanced analytics and integrating machine learning technologies.

- Internet of Things: The Internet is also potentially exposing businesses to security breaches. Increasingly, IoT devices themselves are being turned into Distributed Denial of Services (DDoS) weapons.

- Augmented Reality: Retail companies using virtual stores are applying AR to lead customers away from brick-and-mortar stores to online purchasing. They also enable brick-and-mortar retailers to take virtual showroom experiences to another level that blend digital and physical shopping.

- Automation Consultants: IT teams are developing and tailoring automation software that autonomously frees up staff for more strategic tasks. Some experts, however, predict that AI will soon even take over much of the coding work that AI developers currently do.

- Cybercrimes: Targeted espionage, ransomware, denial of service and privacy breaches present major and rapidly escalating trends that impact IT.[72]

The Most AI-vulnerable Jobs and Careers

Alex Williams asks in a December 11, 2017 *New York Times* article, *Will Robots take Our Children's jobs?* He warns us to give serious thought to that prospect.

Williams points out that while radiologists in New York typically earn about $470,000 per year, a start-up called "Arterys" has a program that can perform a magnetic-resonance imaging analysis of blood flow through a heart in just 15 seconds…compared with 45 minutes required by humans.

He also notes that robots already assist surgeons in removing damaged organs and cancerous tissue. In 2016, a prototype robotic surgeon called "Smart Tissue Autonomous Robot" (STAR) outperformed human surgeons in a test in which both had to repair the severed intestine of a live pig.

According to Williams, any legal job that involves lots of mundane document review (what lawyers spend a lot of time doing) is vulnerable. Software programs are already being used by companies including JPMorgan Chase & Company to scan legal papers and predict what documents are relevant, saving lots of billable hours. Kira Systems, for example, has reportedly cut the time that some lawyers need to review contracts by 20 to 60 percent.

Big banks are now using software programs that can suggest bets, construct hedges and act as robo-economists, using natural language, according to *Bloomberg*. BlackRock, the biggest fund company in the world, has announced that it will replace some highly-paid pickers with computer algorithms.[73]

Process automation can be expected to have an enormous employment impact in many business industries and career fields. Results of a 2015 McKinsey & Company study showed that technologies could automate 45 percent of the activities people are paid to perform and that about 60 percent of all occupations could see 30 percent or more of their constituent activities automated again with technologies [already] available today.[74]

The *McKinsey Quarterly* dated July 20, 2016, lists four key business preconditions for such investments:

- Technical feasibility of automation.
- Cost of developing equipment hardware and software.
- Relative cost and availability of labor being replaced.
- Other benefits beyond labor substitution including higher output, better quality and fewer errors.

McKinsey & Company researchers also describe types of activities that are most susceptible to automation. They attribute the greatest potential to physical tasks in predictable environments, such as operating machinery and preparing fast food. Two other important categories are collection and processing of data, which can be done faster with machines. These, they contend, could displace large amounts of labor—for instance, in mortgage origination, paralegal work, accounting, and back-office transaction processing.[75]

The McKinsey report found that since predictable physical activities figure prominently in sectors such as manufacturing, food service and accommodations and retailing, these are the most susceptible to automation based on technical considerations alone.

Accommodations for food service (first most automatable) include: preparing, cooking or serving food; cleaning food-preparation areas; preparing hot and cold beverages; and collecting dirty dishes. An estimated 73 percent of the activities workers perform in food service and accommodations have the potential for automation, based upon technical considerations.

Manufacturing (second most automatable) involves performing physical activities or operating machinery in a predictable environment. These activities represent one-third of the workers' overall time. Activities range from packaging products to loading materials on production equipment to welding to maintaining equipment.

Within manufacturing, 90 percent of what welders, cutters, solderers and brazers do, for example, has the technical potential for automation.

Retailing: Based upon technical potential, researchers estimate that about 53 percent of retailing activities are automatable. Examples include: technology-driven stock management and logistics, packaging objects for shipping, stocking merchandise, maintaining records of sales, gathering customer or product information and other data-collection activities. Surveyors calculate that 47 percent of a retail salesperson's activities have the technical potential to be automated.

Financial services and insurance: Researchers estimate a technical potential to automate up to 43 percent. Mortgage brokers spend as much as 90 percent of their time processing applications.

More sophisticated verification processes for documents and credit applications could reduce that proportion to about 60 percent. The McKinsey & Company researchers concluded:

Automation will have a lesser effect on jobs that

> *involve managing people, applying expertise, and social interactions, where machines are unable to match human performance for now.*

Automation will also have less impact upon jobs in unpredictable environments—occupations such as gardeners, plumbers, or providers of child-and eldercare. Not only are they technically more difficult to automate, but they are economically less attractive from a business perspective because they typically command lower wages.

On the plus employment side, the researchers conclude:

> *Workers displaced by automation are easily identified, while new jobs that are created from technology are less visible and spread across different sectors and geographies.*

Among the most difficult activities to automate with current technologies are those that involve managing and developing people or that apply expertise to decision making, planning, or creative work. Often characterized as knowledge work, this type of work can be as varied as coding software, creating menus or writing promotional materials.[76]

Categories listed with highest percentage of job growth net of automation include:

- Healthcare workers

- Professionals such as engineers, scientists, accountants, and analysts

- IT professionals and other technology specialists

- Managers and executives, whose work cannot easily be replaced by machines

- Educators, especially in emerging economies with young populations

- "Creatives," a small but growing category of artists, performers and entertainers who will be in demand as rising incomes create more demand for leisure and recreation

- Builders and related professions, particularly in the scenario that involves higher investments in infrastructures and buildings

- Manual and service jobs in unpredictable environments, such as home-health aides and gardeners

The McKinsey & Company's 2017 report concludes that employment transitions to automation will be very challenging—"matching, or even exceeding, the scale of shifts out of agriculture and manufacturing seen in the past:"

> *We previously found that about half of the activities people are paid to do globally could theoretically be automated using currently demonstrated technologies. Very few occupations—less than 5 percent—consist of activities that can be fully automated. However, in about 60 percent of occupations, at least one-third of the constituent activities could be automated, implying substantial workplace transformations and changes for all workers.*[77]

A Second Machine Age

The emergence of an AI revolution in what is chronicled as a "second machine age" is exemplified by rapid advancements and influences of industrial 3-D printing technologies. No longer limited to preliminary equipment prototyping and rapid tooling, it is now serving endlessly diverse manufacturing applications, including the creation of entire jet engines, medical and dental devices, lenses for light-emitting diodes (LEDs), and home appliance parts to name a tiny sampling.

More and more companies are joining the revolution as the range of printable materials continues to expand. In addition to basic plastics and photosensitive resins, they already include ceramics, cement, glass, numerous metals and new thermoplastic composites infused with carbon nanotubes and fibers.

The technology fabricates an object layer-by-layer according to a digital "blueprint" downloaded to a printer which allows not only for limitless customization, but also for designs of great intricacy.

That being said, traditional injection-molding presses can spit out thousands of widgets an hour. Three-D processes are slower but catching up. While still slower, since each unit is built independently, they can easily be modified to meet unique needs.

Writing in the *Harvard Business Review*, Richard D'Aveni notes that a big 3-D printing advantage is that pieces that used to be molded separately and then assembled can be produced as one piece in a single run. D'Aveni cites a simple example of sunglasses:

> *The 3-D process allows the porosity and mixture of plastics to vary in different areas of the frame. The earpieces come out soft and flexible, while the rims holding the lenses are hard. No assembly required.*[78]

Richard D'Aveni further explains that printing parts and products also allows them to be designed with more complex architectures, such as honeycombing within steel panels or geometries previously too fine to mill. Complex mechanical parts—an encased set of gears, for example—can be made without assembly.

Additive methods can be used to combine parts and generate far more interior detailing in products...such as jet engines. GE Aviation uses the technology to create a nozzle that used to be assembled from 20 separately cast parts. Now, it's fabricated in one piece in order to cut costs of manufacturing by 75 percent.

What's more, 3-D printing applications are tackling complex large-scale construction challenges which have previously involved a variety of highly specialized technology operations. So-called "big area additive manufacturing" makes it possible to create endo- and exoskeletons of entire jet fighters, including the body, wings, internal structural panels, embedded wiring and antennas and soon, the entire central load-bearing structure. This makes use of a huge gantry with computerized controls to move the printers into position.

To borrow from an aphorism...the sky is no limit for 3-D printing.

AI Impacts on Healthcare

Dr. Bhardwaj, co-founder and CEO of Innoplexus AG, believes that as AI continues to advance, it has the potential to transform the future of healthcare in three critical areas: advanced computation, statistical analysis and hypothesis generation. Writing in *Forbes.com*, he refers to these general stages advancements in three waves of technology development and application:[79]

- The First-Wave AI: is composed of "knowledge engineering" technology and optimization programs which found efficient solutions to real-world problems...like using AI to estimate a patient's heart disease risk.

- The Second-Wave AI: is characterized by machine learning programs which utilized statistical probability analysis to conduct advanced pattern recognition. In contrast with first-wave AI, second-wave perceives and learns—sometimes more effectively than humans. Yet as Dr. Bhardwaj points out, "These programs still cannot fully replace human assessment because they have not matched humans' capacity for deep data interpretation."

- The Third-Wave AI: This involves emerging programs which can generate novel hypotheses. These technologies are capable of examining huge data sets, identifying statistical patterns, and creating algorithms to explain the patterns. Although they still have a long way to go, these systems have enormously exciting futures.

The enormous potential of third-wave AI programs lies in their ability to increase the quantity of data that can be meaningfully analyzed. The programs identify connections between previously unassociated data points by normalizing the contexts of various points to allow simultaneous generation and testing of novel hypotheses in a host of healthcare scenarios. As a result, they can both learn from and explain complex statistical patterns in order to then teach humans what is learned.

IBM's "Watson for Health" is applying cognitive technology to unlock vast amounts of health data. The program can reportedly review and store every medical journal, symptom and case study of treatment and response around the world—doing so exponentially faster than any human.

Google's "DeepMind Health" is working in partnership with clinicians, researchers and patients to solve healthcare problems. The technology is described to combine machine learning and systems neuroscience to build powerful general-purpose learning algorithms into neural networks that mimic the human brain.

Physician and Stanford University Professor Dr. Robert Pearl reports that today's most common AI uses are in algorithmic evidence-based approaches which are programmed by researchers and clinicians. Writing in *Forbes.com*, he cites key examples:[80]

- Cancer Treatment Protocols: Using consensus algorithms from experts in the field, along with the data that oncologists enter into the medical record (i.e., a patient's age, genetics, cancer staging and associated medical problems), a computer can review dozens, sometimes hundreds, of established treatment alternatives and recommend the most appropriate combination of chemotherapy drugs for a patient.

- Cancer Screening: Computer programs excel at human pattern recognition tasks where the human eye fails even in the best clinicians. Independent studies have found that 50 percent to 63 percent of women who get regular mammograms over 10 years will receive at least one "false-positive" (a test result that is wrong indicates the possibility of cancer, thus requiring additional testing, and sometimes, unnecessary procedures). As many as one-third of the time, two or three radiologists looking at the same mammography will disagree on their interpretation of the results. Visual pattern recognition software, which can store tens of thousands of images, is estimated to be 5 percent to 10 percent more accurate than the average physician.

- Diagnostics: Powerful deep-learning AI applications are

advancing in such diagnostic fields as radiology (CT, MRI and mammography diagnoses), pathology (microscopic and cytological diagnoses), dermatology (rash identification and pigmented lesion evaluation for potential melanoma), and ophthalmology (retinal vessel examination to predict the risk for diabetic retinopathy and cardiovascular disease).

AI-related technologies are finding life-extending and life-saving uses in the general public realm as well. A proliferation of consumer wearables and other medical devices combined with AI are being applied to oversee early-stage heart disease, enabling doctors and other caregivers to better monitor and detect potentially life-threatening episodes at earlier, more treatable stages.

Such advancements offer both the incentives and means to become more deeply involved and interested in our own health, especially when we can easily share these data sets with our health practitioners in a more accurate and structured way. As Dr. Pearl predicts:

> *Over time, patients will be able to use a variety of AI tools to care for themselves, just as they manage so many other aspects of their lives today. It may not happen soon…But sometime in the future—more years than entrepreneurs would like and fewer years than most doctors hope—AI will disrupt healthcare as we know it. Of that we can be sure.*

Pearl adds:

> *Unfortunately, the biggest barrier to artificial intelligence in medicine isn't mathematics. Rather, it's a medical culture that values doctor intuition over*

> evidence-based solutions. Physicians cling to their independence and hate being told what to do. Getting them comfortable with the idea of a machine looking over their shoulder as they practice will prove very difficult for years to come.

On a more hopeful note...for medical professionals, Pearl concludes:

> Without question, the role of the physician will change in the future. Fortunately for doctors, however, computers have yet to demonstrate the kind of empathy and compassion that millions of patients rely on in their medical care.

AI Impacts on the Practice of Law

In his book *Tomorrow's Lawyers: An Introduction to Your Future*, Oxford University Law Professor Richard Susskind identifies three primary drivers of change that will challenge the future of the market for legal services: the "more-for-less" challenge, liberalization in business structures and information technology. He regards this as perhaps the most misunderstood and under-appreciated catalyst of change in service delivery.[81]

Susskind argues that while many lawyers believe that information technology is overhyped, this perspective misses the larger trend exemplified by the persistence of Moore's law: the astounding growth of accessible digital information.

Echoing much of Susskind's point of view, Mark McKamey believes that AI advancements will inevitably influence law practice due to a complimentary compatibility and fit with human abilities. He emphasizes that computers can be programmed to reason at a high-level relatively easily, but struggle with low-level sensorimotor tasks.[82]

Writing in the *Appeal Law Journal*, McKamey observes that "e-discovery" software can already sift through enormous sets of documents to help determine their potential relevance to a special case...during processes of discovery, for example.

Citing a study by Maura Grossman and Gordon Cormack, McKamey notes:

> [T]echnology-assisted review can (and does) yield more accurate results than exhaustive manual review, with much lower effort." Author says that other articles further emphasize the cost benefits of e-discovery, which can amount to savings of 70 percent or more.

Emerging AI systems can also offer legal opinions...for example, advice to a client regarding what the value of a particular personal injury claim might be.

None of this suggests that human judgment will become irrelevant. While lawyers and paralegal assistants will spend far less time sifting through documents, they will continue to remain indispensable to the process.

Mark McKamey points out that law is messy, making it difficult to construct algorithms that capture the law in a useful way. Few legal problems have clear yes or no answers. Legal reasoning is inherently a "parallel process" in which the answer to one question may change which questions are subsequently asked...a problem that can significantly disrupt the ability of computers to deliver useful answers to legal questions.

Citing Evgeny Morosov's book *To Save Everything, Click Here: Technology, Utopianism, and the Urge to Fix Problems that Don't Exist*, McKamey warns about a worrisome "solutionism" trend that results from excessive computer dependence:

> *Solutionism is a kind of technological*

> determinism...the technological solutions available for minor problems...lead us to shallow thinking, and our goals divert from understanding large, complex social problems into writing yet more apps. Worse, we start seeing only problems that can be solved by apps as problems worth solving.

McKamey is particularly concerned that solutionism can creep into new legal technology applications with a commercial focus that obscures justice as the ultimate goal of the legal system.

At the same time, new user-friendly applications are also increasing public access to justice by enabling clients to solve their own problems without consulting legal experts. A boom of legal outside disruption innovation companies is already underway. LegalZoom and Rocket Lawyer, which began by servicing the low-margin end of the market, are ramping their way up and now competing with traditional firms that had abandoned those markets.

Outside pressure from non-traditional legal service providers now leaves traditional practitioners little choice but to embrace technological change more fully. As Richard Susskind submits:

> [It] is simply inconceivable that information technology will radically alter all corners of our economy and society and yet somehow legal work will be exempt from any change.

Adapting to Disruptive AI Influences

In his *Forbes.com* article *Why Robots Will Not Take Over Human Jobs*, business consultant and writer Andrew Arnold emphasizes that, above all, those entering the workforce today will have to be adaptable:

> *They'll have to be hungry for knowledge and committed to continuing education whether that's by taking an online MBA, attending conferences, reading books, consuming podcasts or taking advanced degrees.*[83]

He adds:

> *Workers will need to develop technical skills and keep those skills updated as technology moves forward. Those who do not want to deal with technology need to pursue careers where it is not as much a factor or where demand for human skills and talents remains high.*

McKinsey & Company researchers advise that the share of the workforce that will need to learn adaptive skills essential to find work in impacted occupations is particularly urgent in regions with advanced economies. Their 2017 report projects that up to one-third of the estimated 2030 workforce in the United States and Germany will be impacted, and nearly half in Japan. There is a higher automation potential in Japan than the United States because the weight of sectors that are highly automatable, such as manufacturing, is higher.[84]

The McKinsey team warns that the largest national challenge will be to ensure that there are sufficient numbers of workers with appropriate skills necessary to transition to new job opportunities and needs:

> *Countries that fail to manage this transition could see rising unemployment and depressed wages.*

The researchers emphasize that providing job retraining and enabling individuals to learn marketable new skills throughout their lifetime will be a critical challenge—and for some countries, the central challenge.

Mid-career retraining will become ever more important as the skill mix needed for a successful career changes. Businesses can take the lead in some areas, including with on-the-job training and providing opportunities to workers to upgrade their skills.

Individuals must also face a harsh reality that the idea of a "job for life" is becoming passé. Success and survival will require a continuing need to update and hone skills for jobs of the moment.

As always, there will be winners and losers.

As quoted in *The Guardian*, Dan Collier, chief executive of Elevate said:

> We can't all be knowledge workers. So there will be a lot of unemployment—and perhaps no impetus to help these people. There will end up being a division between the few jobs that need humans, and those that can be automated.[85]

Dave Coplin, chief envisioning officer for Microsoft UK, predicts that this will naturally lead to a divide between high-level, leadership roles and then less highly-specialized jobs that can be automated:

> Even if we can automate all the services we need (and thus eliminate most jobs), we will continue to have huge societal problems that need attention...We are on a burning platform—a key issue of the future will be how will we feed everyone?

Richard Newton, author of *The End of Nice*, pessimistically said that automation is either going to be very good or very bad—

and that either way there's not going to be much in the way of work:

> The defining factor to whether there will be a two-tier society of mass unemployment, or a society of leisure, will be what society places value on...The social contract of work has been ripped up, and people will be left with nothing for as long as businesses and corporations value productivity.

Julia Lindsay, chief executive of the iOpener Institute, predicts that the future workforce is more likely to shift towards more part-time, freelance-based work:

> Employers won't think in terms of employees—they'll think in terms of specialisms. Who do I need? And for how long? Future work may also be focused around making complex decisions—using creativity, leadership and high degrees of self-management.

Mark Spelman, co-head of future of the internet interactive and member of the executive committee of the World Economic Forum, told *The Guardian* that the cheapest and most productive thing to do will be to automate the workforce, so if productivity is what shareholders place value on, there will be mass unemployment:

> But if you use technology to reduce accidents, produce food for people and save time—that provides a great social value...So in future we need to put societal and shareholder value together.

Despite fearful naysayers and doomspeakers, human history demonstrates that we are a very adaptable lot. After all, we not only survived first Industrial Revolution, but came out of it living better and longer than all preceding generations. This new Information Revolution is already expanding these benefits, while dramatically advancing human mental and physical capacities in the process.

There will always be a demand for adaptable and versatile human labor, but workers everywhere will need to rethink traditional notions of where they work, how they work and what talents and capabilities they bring to that work.

Nor is there is any reason to believe that artificial intelligence will replace the need for creative thinking, problem-solving, leadership, teamwork and personal initiative. Far more likely, we humans can leverage technology to provide a better world.

All of this will require fundamentally resetting our collective intuition regarding how we…humanity, can reinvent better versions of ourselves that fit that better world we hope for our children…and for future generations beyond theirs, ad infinitum.

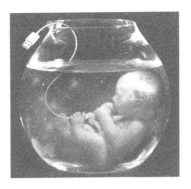

Mechanizing Our Lifestyles: Utopian Resorts or Ant Farms?

THERE IS NO turning back the clock or holding back the advances on the myriad of ways that information technology—AI and Internet connectivity in particular—are changing not only our lifestyles, but our fundamental perceptions regarding the types of lifestyles we deem most desirable as well.

Enthusiastic proponents promise tantalizingly optimistic visions: daily new conveniences previously conceivable in the fertile imaginations of a fiction writer but decades, or even a few years or days, ago; personal living efficiencies and household economies that save precious time and money; enhanced mobility through shared on-demand transportation services that banish most private automobiles to rusty scrap heaps of oblivion; and safety from predatory behaviors of others through ubiquitous, ever-watchful interconnected security devices.

If desired, those same technologies afford opportunities for many of us to live and work pretty much wherever and whenever we prefer through digital telecommuting that connects us to clients and employers in an inherently space-less world of increasingly faceless virtual relationships. In all cases, we remain connected and dependent upon ever watchful and tirelessly obedient digital assistants, who, in turn, monitor, record and market private information we generously, often unwittingly, share.

So on balance, how can each of us assess the ultimate cost-benefit tradeoffs between conveniences and economies and the inevitable encroachments upon privacy and independence? In what ways are these accelerating developments transforming personal, business and societal cultures? Should we be compliant and adaptive, and even positively hopeful—or rather…well actually, there really is no constructive "rather."

Just as with every other major evolutionary game changer, and this is clearly a formatively consequential one, let's get used to the idea. Maybe it will turn out just fine after all.

Influencing Where and How We Live

Telecommuting to work through communication links rather than through physical presence now enables a new breed of entrepreneurs and employees to conduct business in sweatpants and shorts from geographically-dispersed satellite offices, personal homes and vacation retreats. This growing trend represents a substantial departure from a traditional urban model where employment is predominately concentrated in a population center such as a major city's Central Business District (CBD).

According to a 2017 Gallup survey, more Americans are not only working remotely, but they are also doing so for longer periods. Forty-three percent of the more than 15,000 adults surveyed said that they spent at least some time doing so, a four percent increase since a previous 2012 poll.

The share who said they spent a day or less a week working remotely shrank substantially from 2012 to 2016, falling to 25 percent from 34 percent. At the same time, the share that reported working remotely four to five days a week grew by nearly the same amount, rising to 31 percent from 24 percent.[86]

Many business employees and employers have recognized important benefits which continue to encourage remote work and associated flexible scheduling practices. Gallup found that such opportunities often played a major role in an employee's decision to take or leave a job. The report said:

> *Employees are pushing companies to break down the long-established structures and policies that traditionally have influenced their workdays.*

Although widespread and on the rise across all industries, some types of businesses and organizations reported better successes with remote working arrangements than others. The concept was particularly popular in finance, insurance, real estate, transportation, manufacturing and construction, and retail industries where those who reported working remotely rose eight percentage points to 47 percent from 2012 to 2016. There were also steady gains in healthcare, law and public policy.

Well over half of all employees in transportation, computer, information systems and mathematics fields worked remotely some of the time. However, remote work had become less common than in 2012 for Americans employed in the fields of community and social services; science, engineering and architecture; and education, training and libraries.

While both employees and some employers viewed remote work arrangements to be broadly beneficial, even those representing popular industry applications reported struggling with the extent it should be embraced. For example, Yahoo and the insurance giant Aetna, which had originally pioneered such plans, had second thoughts regarding negative influences upon collaborative engagement.

Perhaps surprisingly, in 2012 the workers who said they felt most engaged with the employing organization while working remotely were those who spent the least amount of time off-site. By 2016, that was no longer true…those who spent none or all of their time out of the office reported feeling equally engaged. In fact, those who spent 60 percent to 80 percent of their time away from the office expressed the highest rates of engagement. Gallup reported:

> In spite of the additional time away from managers and co-workers, they are the most likely of all employees to strongly agree that someone at work cares about them as a person, encourages their development and has talked to them about their progress.

It is natural to expect remote working arrangements to cause employing business cultures to become far more impersonal as places where people don't really get to know each other: who, for example, is having a personal problem or celebrating a birthday? Employees don't meet the boss in the elevator. They don't discuss casual ideas that give rise to important opportunities. They don't share corporate or individual achievements. They become environments where "every man (and woman) is an island."

Counterintuitively, Gallup reported that those who spent three or four days a week working remotely were also the most likely to report thinking that they had a best friend at work, and had opportunities for professional growth.

Writing in the Harvard Business Review, Sean Graber posits "why remote work thrives in some companies and fails in others." He explains:

> Successful remote work is based on three core principles: communication, coordination, and culture. Broadly speaking, communication is the ability to

> *exchange information, coordination is the ability to work around a common goal, and culture is a shared set of customs that foster trust and engagement.*[87]

Graber points out a special challenge to success is that communication in a virtual environment can make it more difficult to explain complex ideas. This results when a lack of face-to-face interaction limits social cues which may often lead to misunderstandings and potential conflicts. This can be particularly true when people aren't able to ask questions and carry on discussions in real time.

Sean Graber also warns that failures of opportunities to meet face-to-face can also tend to degrade and compromise a company's social and working culture. This can occur, for example, when out-of-office individuals become inclined to focus on their individual task assignments at the expense of team spirit and collaboration.

Here again, information technology is reconnecting and extending virtual face-to-face teleconferencing opportunities across boundless business, professional and social landscapes. Even small offices and organizations with no offices at all now have access to high-speed Internet and cheap web cameras, headsets, speakers and even smartphones.

Video teleconferencing is becoming enormously popular for a variety of reasons. It affords a means to host meetings from locations anywhere: to visually and verbally communicate with dispersed audiences in real-time; to discuss and resolve time-critical matters; to share, exchange, create and approve documents; to record, save and re-distribute proceedings in text and video formats; and to accomplish all of this on short notice and without incurring travel and lodging expenses.

Organizations ranging from large corporations to small start-ups can also mutually enjoy substantial benefits afforded through remote tele-employment. Key among these, the strategy enables them to recruit and retain the best people no matter where they live, to buy a unit of service and labor at lower salary and overhead prices and to minimize personnel facility requirements.

Forbes contributor William Arruda believes that the biggest winners in flexible work programs are the company employees. Notwithstanding the fact that the Internet connects businesses to specialized freelance solopreneurs who compete for their work projects and jobs, businesses operations in general will increasingly decentralize.

Arruda writes:

> [T]he new trend that's exciting me and is growing exponentially is the area of remote work that's not for freelancers; it's for employees. There's no arguing that the 9-to-5, 40-hour work week, with your entire team located near you, is gone. And it's not coming back. Today, it is more likely that you work on a team where some or even all of your colleagues work remotely.[88]

Also writing in *Forbes.com*, Tiffany Williams enthusiastically foresees enormous business networking and service market opportunities for new generations of work-from-home "digital nomads" who are now enabled to strike individually self-determined balances between their personal and professional lives. Practitioners working from mobile virtual offices have become free to travel, spend more quality time with families and friends, pursue educational goals and special hobbies and attend to medical issues.[89]

Internet connectivity now instantly hooks up anyone with a laptop or smartphone to markets and products throughout the world. Anything from accounting, legal transcription, writing and marketing can be accomplished remotely. Advanced information processing software which was previously used only by corporate enterprises due to lack of affordability can now access this through the cloud.

Williams observes the rapidly growing popularity of commission-based product and service marketing businesses strategies which are ideal for home businesses. Online networking enables businesses to concentrate only on producing and delivering great products and services and outsource promotion to outside professionals who are compensated according to sales performance.

Home-connected businesses afford working parents and students flexibility to arrange schedules around other priorities. Living close to their children's schools and avoiding long work commutes enables more time to spend with them and less money to spend on day care costs. Online teaching and training programs enable home-based business parents and others to advance learning goals such as attainment of additional knowledge-based skills and credentials.

Andrea Loubier emphasizes that telecommuting benefits, whether through arrangements with primary employers or solo entrepreneurship from homes or close-by neighborhood offices, are about much more than just enabling happy campers to conduct business from Starbucks or beach blankets. Writing in *Forbes.com*, she observes:

> *It allows workers to retain more of their time in the day and adjust to their personal mental and physical well-being needs that optimize productivity. Removing something as simple as a twenty minute commute to work can make all world of difference. If you are ill, telecommuting allows one to recover faster without being forced to be in the office.*[90]

Loubier is enthusiastic about benefits to corporate business employers as well. For example, it can increase employee performance and productivity by reducing distractions such as water cooler gossip. Telecommuting arrangements can also dramatically reduce business operating costs. She writes:

> According to Aetna, an insurance giant in America, it shed 2.7 million square feet of office space and as a result saved $78 million. American Express reported similar results by saving $10-15 million annually thanks to its telecommuting policies.

Telecommuters can save big on cost of living benefits as well. No longer required to locate near a teeming metropolis, home buyers can escape having to pay exorbitant housing prices to live in densely-packed cities and near-by high-priced suburbs. Telecommuters are no longer required to endure daily hours on congested traffic lanes which can be spent in more satisfying and productive pursuits.

Telecommuters can select home sites that offer special cultural, economic, educational and natural amenities which are most important to them and their families: locations which are safest and which offer the best schools for their children; places near other loved ones, including aging parents; settings with access to convenient retail services, and where they can enjoy treasured outdoor landscapes and activities; and communities where people make it a point to know neighbors and to engage long-term friendships.

Just How Smart are Smart Cities?

So okay, technology-enabled remote work opportunities are attracting more and more people to people to rural and "small town" community lifestyles. But what about technological influences on cities? Lots of people prefer living in them too, and there is every reason to believe that many more will continue to do so.

Won't technology make them better, more livable, more vital and prosperous, more convenient and efficient...more "smart"?

Well maybe. After all, we hear more and more that it will.

But for starters, just what, exactly, does the term "smart city" mean?

As it turns out, there are multiple definitions, depending upon who we ask.

The concept of smart cities became a buzzword for both developing new cities for future growth and upgrading existing ones. The idea isn't entirely new. It arguably dates back to the invention of automated traffic lights first deployed in 1922 in Houston and has morphed and crystallized into an image of the city as a vast, efficient robot—a dream of giant technology companies.

The modern movement stems from the idea of digital information and communication technologies (ICT) which obtain large sets of data which are transformed and applied to urban policies.

There are two dominant smart city perceptions. The first and most prevalent is one where the urban fabric and everyone in it becomes increasingly instrumented and ubiquitously monitored and controlled from "everywhere." In this case, technology is typically viewed as a primary driver of change rather than being relegated to only serving as but one means to move the cities higher on the development ladder. In this new language of "smartness," the emphasis is upon smart information, smart meters, smart grids and smart buildings which are rapidly becoming ever-larger parts of our everyday lives.

The second perception envisions the smart city's economy driven by innovation and entrepreneurship with the goal of attracting business and jobs and focusing on efficiency, savings, productivity and competitiveness. ICT's role here is to facilitate and streamline private and public initiatives. The idea of smartness here focuses on smart use of resources, smart and effective management, and smart social inclusion. Within this view, the ICTs are one component of the concept, but by no means its bread and butter.

There is a dichotomy in this picture: corporate utopian visions ("ICT will save us") vs. an academic circle definition which is more varied, diverse and complex. Nevertheless, the two perspectives have morphed to become applied interchangeably, an ambiguous and not always positively connoted mix of automated utopian fantasy lands overseen by robotic versions of George Orwell's Big Brother.

"Smart cities" have come to be marketed as a conveniently imprecise, rhetorical and ideological rationalism for technology reigning supreme over everything...a panacea for all urban ills. Promoters market ICT solutions as quick and effective ways to deal with all manner of urban problems: growing populations, climate change, environmental shocks and other urban threats, including local urban crime problems, congestion, inefficient services and even economic stagnation.

Such rhetoric energetically promulgated by big technology, engineering and consulting companies is predicated on the embedding of computerized sensors into the urban fabric so that bike racks and lamp posts, CCTV and traffic lights, remote-control air conditioning systems and home appliances all become interconnected into the wireless broadband Internet of Things.

Writing in *The Guardian*, contributor Steven Poole asks:

> And what role will the citizen play? That of unpaid data-clerk, voluntarily contributing information to an urban database that is monetized by private companies? Is the city dweller visualized as a

smoothly moving pixel, travelling to work, shops and home again, on a colorful 3-D graphic display? Or is the citizen rightfully an unpredictable source of obstreperous demands and assertions of rights? [91]

Critic Bruce Sterling says "stop saying smart cities." He wrote in The Atlantic:

> *The digital techniques that smart-city fans adore are flimsy and flashy—and some are even pernicious—but they absolutely will be used in cities. They already have an urban heritage. When you bury fiber-optic under the curbs around the town, then you get Internet. When you have towers and smartphones, then you get ubiquity. When you break up smartphones into separate sensors, switches, and little radios, then you get the Internet of Things.*
>
> *These tedious yet important digital transformations have been creeping into town for a couple of generations. At this point, they're pretty much all that urban populations can remember how to do. Google, Apple, Facebook, Amazon, Baidu, Alibaba, Tencent—these are the true industrial titans of our era. That's how people make money, that's how people make war, so of course, it will be how they make cities.*
>
> *However, the cities of the future won't be 'smart,', or well-engineered, cleverly designed, just, clean, fair, green, sustainable, safe, healthy, affordable, or resilient. They won't have any particularly higher ethical values of liberty, equality, or fraternity, either. The future city will be the internet, the mobile*

> cloud. And a lot of weird paste-on gadgetry by City Hall, mostly for the sake of making towns more attractive to capital.[92]

Whole new cities, such as Songdo in South Korea, have already been constructed according to this template. Songdo's buildings have automatic climate control and computerized access; its roads, water, waste and electricity systems are dense with electronic sensors to enable the city's brain to track and respond to the movement of residents.

In India, Prime Minister Narendra Modi has promised to build at least 100 smart cities. Dholera, the first, is currently an empty backdrop landscape-in-waiting due in large part to flooding problems which have discouraged private investment. Despite promotional arguments that the city was planned as a green field development to serve the "public good," peasant and farmer protestors who lost their lands angrily disagree.[93]

One of the most ambitious pilot projects of this kind was in Rio de Janeiro and involved constructing a large state-of-art Center of Operations operated by IBM. Branded as "Smarter Planet," it featured impressive control center dashboards to provide panoptical views of huge amounts data.

Rio made the great investment into the smart system as a tool to predict and manage a flood response in advance of two huge events—the FIFA World Cup in 2014, and the Olympic Games in 2016. That purpose soon morphed into an enormously larger urban surveillance information gathering and processing enterprise. Quoting Rio's mayor, Eduardo Paes:

> The operations center allows us to have people looking into every corner of the city, 24 hours a day, seven days a week.[94]

As Steven Poole warns:

> The things that enable that approach—a vast network of sensors amounting to millions of electronic ears, eyes and noses—also potentially enable the future city to be a vast arena of perfect and permanent surveillance by whomever has access to the data feeds.[95]

The project has also raised many practical questions regarding whether and how it improved the life of its citizens and in what ways it made the city smart.

Urban researchers Milan Husar, Vladimir Oudrejieka and Sila Ceren Varis point out that Rio's smart planning demonstrated a variety of inherent modeling problems which bring together certain distinct aspects which do not always fit together, while simultaneously hiding other issues. For example, modeling of weather predictions for events such as floods and traffic congestion used hypothetical statistical probabilities instead of descriptive certainties which can never be known in advance.

The best that can be done is to produce a set of possible scenarios while accepting and accounting for inherent uncertainties. The authors write:

> In practice, this means that city representatives can only use these scenarios as aids for decision making on the basis of personal, electoral or financial risk if to take preventative action or not.[96]

A major risk of this model-based planning is that city governments become inclined to use the official-appearing "data from a distance" to justify further regulation and control over urban systems and populations. Unlike in case of rational comprehensive planning in the 1970s, the current information technology-intensive approach has a greater chance of achieving political agenda-driven objectives they are inadvertently or intentionally programmed to support.

They also offer cover for bad decisions. City managers can always claim, "It wasn't me that made the decision, it was the data." [97]

Trading Our Privacy for Convenience?

Privacy should be a major concern. Large data companies are already collecting data on their users and until now people had to be connected to their network to be seen. But now, so-called "smart cities" are installing CCTV and other devices into panopticons where people can be watched every moment of their lives.

As marketed and presented, the data and the algorithms processing it are benign and indifferently neutral. We citizens who are constantly being observed and recorded are simply perceived as data points...information generators in various representative nodes in a system designed around the idea of data mining our ant-like patterns of behavior. [98]

Nevertheless, this pervasive, ever-vigilant monitoring of our collective and individual activities and habits portends some very frightening implications regarding relinquishment of our privacy and prerogatives to invisible voyeurs and agenda-driven societal power-brokers who claim to represent our best interests. As Songdo, South Korea researcher S.T. Shawayri points out, the data is never neutral, essential and objective in its nature. It is invariably "cooked" to recipe by chiefs embedded within institutions with aspirations and goals. [99]

And as Steven Poole observes:

> In truth, competing visions of the smart city are proxies for competing visions of society, and in particular about who holds power in society. [100]

Just how worried should we be regarding which vision prevails and who wields what influence over the broad personal and public aspects of our lives? In contemplating your answer, try to imagine the sort of life you would be willing to accept for yourself, or your children and grandchildren a decade or two from now given the trends we are already witnessing today.

Imagine, for example, that as soon as you step outside your door—maybe even before—your actions are swept into a digital dragnet. Video surveillance cameras placed everywhere record footage of your face which becomes entered for correlation by feature recognition algorithms matched with photos on your driver's license to a national ID database.

The reason, of course, is that this is being done in your best interest to enhance your safety.

And imagine that algorithms keep track of what you purchase online; where you are at any given time; who your friends are and how you interact with them; how many hours you spend watching television or playing video games; and what bills you pay (or not).

Actually, this should be quite easy to imagine.

As Rachel Botsman reminds us in her book *Who Can You Trust? How Technology Brought Us Together and Why It Might Drive Us Apart,* most of this already happens thanks to social media data-collecting behemoths. Ms. Botsman informs us that a 2015 Office of Economic Development (OECD) study revealed that at that time there were already about one-quarter as many connected private and government-operated monitoring devices in the United States as the entire population.

Other agencies are likely to follow the National Security Agency (NSA) watchdog lead. Although the U.S. Department of Transportation Security Administration scrapped a proposal to expand PreCheck background checks to include social media records, location data and purchase history following heavy criticism, a major new terror incident can readily revive the plan.[101]

Now further imagine how this seemingly innocuous "technology for the public good" trend might lead to frighteningly bad public and personal consequences far more rapidly than we realize.

In June 2014, the State Council of China published a properly ominous-sounding document called *Planning Outline for the Construction of a Social Credit System*. Although the lengthy policy document appeared rather innocuous to many, the idea was anything but inconsequential.

Every citizen would "earn" a Social Credit System (SCS) score to rank how trustworthy they are. Rankings across the entire 1.3 billion population was pitched as a desirable way to measure and build a culture of "sincerity."

As the document explained:

> It will forge a public opinion environment where keeping trust is glorious. It will strengthen sincerity in government affairs, commercial sincerity, social sincerity and the construction of judicial credibility.[102]

After all, what could possibly go wrong?

More than 70 years ago two Americans, Bill Fair and Earl Isaac, created the Fair Isaac Corporation (FICO), a data analytics company, to establish credit scores applied by companies to determine many financial decisions. These client service rankings include determinations regarding whether an applicant should be given a loan, and what interest rate should be applied to a consumer property mortgage.

So how will the Chinese SCS rankings be rated? That depends upon which of eight licensed competing data mining company candidates comes up with the most satisfactory plan.

One big contender, social network Tencent, developer of

the messaging app *WeChat* with more than 850 million active users, has teamed with partner China Rapid Finance. Combining the functionality of Facebook, iMessage, PayPal, Instagram, Expedia, Skype, WebMD and many others, WeChat hit 300 million users by its second anniversary.

Tencent is making extraordinary new inroads in ramping up AI-based healthcare supported by scores of researchers operating out of a new Seattle, Washington satellite. The company is also rapidly expanding into multimillion dollar U.S.-based drug discovery deals.[103]

Another big Chinese social credit system development competitor is Sesame Credit. Sesame Credit is run by Ant Financial Services Group (AFSG), the world's largest money market fund. In 2017, Ant, an affiliate company of Alibaba, handled more payment transactions than Mastercard. Ant's AliPay, its payments arm, determines who can qualify to purchase things online, including restaurant charges, taxi fares, school fees, cinema tickets, and a variety of other money transfers.

Sesame Credit has teamed with other data-generating platforms such Didi Chuzing, Uber's main competitor, prior to acquiring the American ride-hailing company's Chinese operations, along with Baihe, the country's largest online matchmaking service. Altogether, this gives them access to gargantuan amounts of data that Sesame can tap to assess how people behave and rate them accordingly.[104]

The Sesame plan for China proposes to rate individuals on scores ranging from 350 to 950 points based upon a "complex algorithm" which takes five factors into account: each citizen's credit history; their fulfillment capacity; and personal characteristics, such as verifying mobile phone numbers and physical addresses; their behavior and performance; and their interpersonal relationships.

Under the fourth category in this system, personal behavior

and performance, a person's shopping habits, for example, might be correlated with measurements of character. As Sesame's Technology Director Li Yingyun explains:

> *Someone who plays video games for ten hours a day, for example, would be considered an idle person...Someone who frequently buys diapers would be considered as probably a parent, who on balance is more likely to have a sense of responsibility.*[105]

Sesame's fifth category delves into each individual's choice of online friends and interactions to assess "positive energy." This suggests that nice messages regarding how well the government is performing and the country's economy is doing will make the score go up.

As for posting dissenting political opinions or critical observations—that's probably not a great idea. It's also not very wise to associate with any online friends who do so out of concern for their own score being penalized.[106]

Alibaba, which sells insurance products and provides loans to small-to-medium-sized businesses, is working to become a major player in planning smart cities throughout the world. The company's AI cloud platform, "ET City Brain," develops algorithms to predict outcomes across traffic management, healthcare and urban planning and crunching data from cameras, sensors, social media and government statistics.

Aiming to revolutionize urban management, Alibaba has partnered with Nvidia for its deep-learning-based video platform for smart city services. More recently, Alibaba also led a financing round for Chinese computer vision startup "SenseTime," has backed AI-based vehicle-to-vehicle network developer Nexar and has partnered with the Malaysian government to launch the country's first City Brain initiative to optimize urban traffic flow.

Alibaba is getting deeply involved in monitoring and potentially controlling human traffic as well. Already geared with facial recognition for online retail and mobile user sign-ins, Alibaba's AliPay has more than half a billion users worldwide.

The company is now working with several local Chinese governments, including Macau and Hangzhou, and is expanding its reach far beyond Asia with operations already in more than 200 countries.

Alibaba is set to launch a major high-tech hub in Tel Aviv and at least six other cities. These activities will be supported by a global $15 billion R&D initiative in AI, quantum computing and emerging tech-driven markets.[107]

The Chinese government has adopted a voluntary sign-up trial watch-and-learn experiment approach regarding the social merit approach prior to plans for a national program roll-out in 2020. This being the case, why have millions of people already signed up to be subjects of a government surveillance system?

There are at least a couple of not-necessarily mutually exclusive possibilities.

One is that those who fail to sign up may face real risks of government reprisals. Another is that those who participate have opportunities to earn "trustworthy" points for financial rewards and "special privileges."[108]

Based upon the Sesame scheme, those whose score reach 600 points can take out a "Just Spend" loan of up to 5,000 yuan to shop online…so long as they do so on an Alibaba site. Reach 650 points and they may rent a car without leaving a deposit, get faster check-in at hotels and use VIP check-in at the Beijing Capital Airport. Those with more than 666 points can get a cash loan of up to 50,000 yuan from Ant Financial Services; above 700, qualify for Singapore travel without supporting documents such as an employee letter; and at 750, earn a fast-tracked application to a coveted pan-European Schengen visa.

As for the downside risk for credit score slackers, although

Sesame's plan claims not to directly penalize "untrustworthy" people, that isn't a promise anyone can literally bank on. As Hu Tao, Sesame's chief credit manager warns, the system is designed so that "untrustworthy people can't rent a car, can't borrow money or even can't find a job." She has even disclosed that Sesame Credit has approached China's Education Bureau about sharing a list of students who cheated on national examinations, in order to make them pay into the future for their dishonesty.

On September 25, 2016, the State Council General Office updated its policy entitled "Warning and Punishment Mechanisms for Persons Subject to Enforcement of Trust-Breaking." The policy document directly and unambiguously states:

> *If trust is broken in one place, restrictions are imposed everywhere.*

As Rachel Botsman reports, people with low ratings will have slower Internet speeds; restricted access to restaurants, nightclubs or golf courses; the removal of the right to travel freely abroad; ineligibility to qualify for rental applications, insurance, loans and even social security benefits; lack of access to certain jobs including civil service, journalism and the legal fields; and restrictions to enrolling themselves or their children in private schools.

The government policy document specifically warns of "restrictive control on consumption within holiday areas or travel businesses." It states that the social credit system "will allow the trustworthy to roam anywhere under heaven while making it hard for the discredited to take a single step." [109]

Each of those discredited steps can soon be monitored by China's planned facial recognition surveillance system that will track movements of trustworthy and untrustworthy citizens with equal impunity. Chinese state media report that the system will be able to scan the country's 1.4 billion-strong population with 99.8 percent accuracy.

The country's state-run Xinhua reported that police using a surveillance system in Nanchang, southeastern China, managed to locate and identify a wanted suspect out of 60,000 attendees at a pop concert. A BBC reporter who cooperated with police in testing a facial recognition system in Guiyang, southwest China, found that it took only seven minutes for authorities to track and locate him.[110]

It seems apparent that smart cities have already gained the ability to outsmart everyone.[111]

Will Automated Vehicles Kill Car Romance?

As Eric Risberg nostalgically observes:

> *From the romance of the road trip to the feeling of getting your driver's license, the car has always conjured images of freedom and control. Those time-worn ideals, however, may soon be a relic of the past.*[112]

Risberg reflects that our current idea of the car is based on the individual or family:

> *[I]t is personal travel, single ownership and individualism in general that fuel our car-obsessed culture. But with driverless cars, a large part of transportation and commuting could instead be thought of as an interlinked system where cars are simply moving through our streets, and then only if and when needed. As a result, the currently neat line dividing public and private transit might start to get*

a bit fuzzy.

And after all, isn't this all for the good? Think about the benefits we are promised. We can still have the privacy and effectiveness of a single car for rides to a work or meeting place remote from a metro line. We can check our emails and text on the way (or whatever replaces them—newspapers will be a thing of the past), plus take along our breakfast and leave any spilled coffee and crumbs behind for the next rider to deal with.

We can be dropped off in the crowded city and be free of the vehicle along with parking space for other uses. We will also no longer need to store and maintain a thirty-or-forty thousand dollar machine in our garage.

But what about those of us who are, for any dumb, backward notion, still very attached to the idea of having those beast and beauty relics? Writing in *The Atlantic*, Ian Bogoost reminds us that private automobile ownership symbolizes a core American tradition of access and freedom. He writes:

> *A car made any place accessible, even if just in theory. That's also why cars became a means of self-expression: How to go somewhere was a choice, so the type and color of a car conveyed a style.*[113]

Bogoost then asks:

> *How can a car offer freedom when you have to ask a company to let you use it? Instead, cars will recede into the background. They will become infrastructure—still important, but unseen unless they break down.*
>
> *Nobody will care what anyone drives, no more than they might ponder the manufacturer of elevator cabins*

> or subway rolling stock. It might be annoying when the elevator takes forever or the train doesn't come, but these matters are akin to acts of God, conducted outside ordinary people's influence. And as the intimate familiarity of choosing, operating, and maintaining vehicles recedes, people will develop a new tolerance for whatever the companies that run the services choose, in terms of appearance and access.

In any case, expect to get used to the idea. Supported by smart city technology and ideology, autonomous self-driving vehicles are racing our way. Waymo, Google's autonomous division, Tesla and Uber are currently leading the pack.

Google had been testing autonomous cars in Mountain View for years before moving its autonomous division to California's Central Valley. Uber moved its test fleet to Tempe in 2017 after a dispute with California over permitting. Uber also operates an autonomous test fleet in Pittsburgh where it lured top computer-vision and robotics researchers from Carnegie Mellon to help it transform car services from flex work to automation.

Presently, Uber refers to driverless car drivers as "pilots" or "operators." Their role is to coax the vehicles along much like copilots or rally navigators as computers do the work. The humans hold laptops for visualization of the road ahead which is captured by a LIDAR unit atop the vehicle. A remote-sensing laser is used for guidance, with data processed by the vehicle's onboard computers.[114]

We can fully imagine that there are some technical issues to be resolved before the user public gains full confidence regarding pilotless safety. What happens, for example, when a computer has to make a split second decision to either save the car occupants or a school bus just ahead?

And how can the systems and the operators who control

them prevent hackers and cybercriminals from sabotaging individual vehicles or even entire fleet networks.

Jeff Williams, chief technology officer of the security firm Contrast Security warns:

> *Nobody today designs cars to operate on the internet, but all of a sudden we are connecting them. And so then we are getting thrown in the deep end....When you start adding technologies like Uconnect and all of a sudden your car is connected directly to the Internet and your car's IP address, then you are accessible from any computer in the world. We have networked all these things and now they are remotely attackable.*[115]

Williams adds:

> *If you take something that was designed to work in one set environment and you connect to it a much more hostile environment, you don't have the right defenses in place. So of course it's vulnerable. It's like Bambi walking out of the forest into the field.*
>
> *While researchers have been the primary ones to expose these weaknesses, it's only a matter of time before nefarious hackers catch on and figure out a way to make money off exploiting these dangerous vulnerabilities. Then we could potentially have a big problem on our hands.*

Back in 2015, hackers were able to take control of a Jeep Cherokee, shutting down its engine on a St. Louis highway. The incident prompted Fiat Chrysler to send out USB sticks to their drivers that contained software fixes to prevent the exploit.

In 2017, for the second year in a row, a group of Chinese

hackers was able to demonstrate a software exploit within a Tesla vehicle, allowing them to take control of a Model X's brake system, audio and door systems. Luckily, these hackers were on our side, working to show the vulnerabilities in poorly secured software.[116]

But let's fully assume that autonomous vehicles will demonstrate high levels of safety and efficiencies, and that many people, most particularly those who live in or commute to cities, will benefit. There should be little doubt that their use will relieve commuter and resident parking and vehicle costs, and may ease urban congestion as well.

Writing from Tempe, Uber's primary test city, Ian Bogoost ponders what ways these big-corporation-owned driverless marvels will change not only the customer experiences but urban cultural experiences as well. He urges us to realize that these new technologies are *not* simply tools that help us accomplish existing goals, like getting to the grocery store or commuting to work. They are changing how we think of space, our concepts of urban design and even the ways we relate to the world.

It is impossible for each of us to fully know what it will be like to live with primary dependence on automated taxi services. While they will free some people from certain risks and burdens of transportation, they and others will simultaneously be bound to new burdens when we allow these technology services to take control of cities.

Bogoost contemplates:

> *So much of the American city is taken up by roads filled with cars filled with people. A huge cultural shift will take place if all that space gets ceded to a few technology companies. Not just big things, but small ones too, like standing on medians and waiting at curbs and perceiving of different colors and styles.*

> *Tiny experiences like these frame the contours of daily life in subtle ways.*[117]

In addition to sketching out the technological, ecological, health and civic impacts of self-driving cars, Bogoost ponders what it will be like to live with them when they cease to be uncanny and just become normal elements of the urban texture. Describing his personal observation, he writes:

> [T]he age of autonomous cars has felt abstract and hypothetical, so far—the stuff of splashy corporate demonstrations and tech-guru prognostications, not everyday life…But standing inches away from the robot Uber, I'm hit for the first time by the tangible, ordinary reality of the future. This isn't a test track or a promotional video. This is a self-driving car in the belly of car-loving, suburban America.

Bogoost adds:

> It seems clear that cars will always have a place in America. But until now, they have been slaved to the people who drive them. The roads have always belonged to people, even if those people were assumed to be inside automobiles. When that coupling is broken for good, and everywhere, the roads will likely be safer, cleaner, and more efficient. But the urban experience, especially in cities like this one [Tempe], will change forever.

But then, it will change anyway—and for many inevitably interconnected technological, cultural and economic reasons—just as societies and cities always have.

Delegating Business Decisions to Algorithms

As we increasingly trust AI and the Internet of Things to run our cities, our lives and our movements, what about letting them run our corporate business economies as well?

Actually, we don't have to wait for this to occur. They are already influencing the national stock market, making major strategic decisions in a wide variety of industrial operations and earning huge profits for big data service providers.

As Steven Cohen and Matthew Granade discuss in a *Wall Street Journal* article titled *Models Will Rule the World,* most of today's industry-leading companies are software model-driven companies. Not all of them started out that way...Aptiv and Domino's Pizza, for instance, are longstanding leaders in their sectors that have adopted their proprietary internal computer software to maintain or extend their competitive dominance.[118]

Model-driven businesses primarily operate based on decision framework platforms with logic derived by algorithms from data, rather than being explicitly programmed by a programmer or implicitly conveyed via individual intuition. The outputs provide predictions upon which decisions are made. Once created, a model can learn from its successes and failures with speed and sophistication that humans usually can't match.

Building such systems requires a combination of capabilities and steps: a program (often software-based) to collect desired data, processes to create models from the data, the models themselves and a mechanism (also often software-based) to deliver or act on the suggestions from those models. The predictive results are then applied to power key business decisions, such as ways to create or optimize revenue streams and to improve cost efficiencies.

Those modeling systems never stop learning from previous decision performance lessons. In other words, model-based businesses are far more than simply data-driven businesses. In model-based businesses, the continuously upgraded algorithms play important roles in defining those businesses. In a data-driven business the data helps a business, but in a model-driven business, the models *are* the business.

Advanced software algorithm modeling systems are being created inside incumbents and startups across a range of industries. For example:

- *Agriculture:* Uses models to improve crop yields. Monsanto, for example, uses comprehensive data collection, processing and algorithm assets to produce more resilient crops by predicting optimal places for farmers to plant based upon historical yields and weather data. Tractors equipped with GPS and other sensors along with field data collected from satellite imagery provide information to support estimates of where rainfall will pool, and to take subtle variations in soil chemistry into account.

- *Logistics businesses*: Use models to overcome E-commerce challenges. For example, a typical fulfillment center has human pickers walking 15 miles a day through warehouses to assemble orders. Accordingly, inVia builds robots that can autonomously navigate a warehouse and pull totes from shelves to deliver them to a stationary picker. The model-driven system considers item popularity and its particular probability of association (putting sunglasses near sunscreen, for example) in order to adjust warehouse layout automatically and minimize the miles robots must travel. Every new order provides a universe of prior predictions and improves productivity across the entire system.

- *The services sector:* Labor-intensive industries such as

those providing international marketing services spend many billions of dollars annually translating everything from catalogs to their terms of agreement into hundreds of languages. While Google Translate works well for the everyday customer, businesses typically require the more sophisticated skills of a human translator. Lilt is building software that inserts a model into the middle of the process. Instead of working from only the original text, the software is presented with a set of suggestions from the model to refine as needed. The model constantly learns from changes the translators make, simultaneously making all the other translators more productive in future projects.

Some of the smartest and most economically impactful computer business models rule decisions in corner offices and trading floors on Wall Street.

The days when traders bought everything from stocks and bonds to bundled credit card debt and oil based upon information tips and instincts are largely gone. That was a time when Morgan Stanley chief John Mack would tell his traders "There's blood in the water. Let's go kill." [2] They did, and were richly rewarded for it.

Today, much of that market warfare is being accomplished by clever software with snippets of code. Algorithms monitor and predict who and what is moving the market, which companies are best positioned to win and lose and which are most desperate and vulnerable for a profitable takeover. They keep tabs on the banks' positions, value complex instruments known as derivatives, generate price quotes for clients, match buyers and sellers and flag unseen risks. They can predict how a drop in the U.S. dollar might affect corn prices, or how to value

---

[2] *Trading Places,* Liz Hoffman and Telis Demos, August 18, 2018, Wall Street Journal

a loan to Ferrari if the Italian government raises interest rates.

As Liz Hoffman and Telis Demos explain in their *Wall Street Journal* article *Trading Places,* the rise of automation occurred partly in response to trading scandals following a 2008 market meltdown. The financial crisis encouraged banks to take discretion away from error-prone and ego-driven humans. We seem to be heading to a point when our entire U.S. economy is being controlled by bots.[119]

But those algorithms, smart as are, haven't proven infallible so far either. A May 6, 2010 "flash crash" caused by algorithms caused prices of U.S. stock market shares to fall 6 percent in 5 minutes.

Although financial turbulence always has and always will be a market risk, computer algorithmic-driven trading strategies are controlling asset prices as never before.

According to Steen Jakobsen, chief investment officer at Saxo Bank, there's a simple rule of thumb that as long as the market isn't down more than 5-6 percent over two-three days: "everything will be fine and volatility will tail off." But a fall more than 6-7 percent on consecutive closes means everyone has to scramble for cover and protection.

Jacobsen warns that this is a big source of danger. The "algos," risk parity funds, ETFs, commodity trading advisors (CTAs) and volatility strategies are all positioned the same way. In many ways, the biggest risk to the market is the market itself.[120]

It can be even more risky when those algorithms are not positioned the same way—conditions when they are designed to deceive or fight one another into making bad decisions.

Trading algorithms like the Volume-Weighted Average Price (VWAP) allow companies to buy massive volumes of shares in lots of small chunks so that other traders aren't tipped off to hone in on the action with their own buys that drive up the price.[121]

Some warring algorithms are programmed to actually prey on other algorithms. Through a process of "algo-sniffing" they are designed to detect the signature of a big VWAP in order to purchase a particular corporation's shares—then buy them faster than the VWAP and sell them to the VWAP at a profit. Although VWAP users condemn algo-sniffing as unfair, it remains legal. This has led to a growing business in adding "anti-gaming" features to execution algorithms to make their use more difficult.

Another even more devious set of strategies work to fool other algorithms through "layering" or "spoofing." A spoofer, for instance, might buy a block of shares and then issue a large number of buy orders for the same shares at prices just fractions below the current market price. This then fools other algorithms and human traders into seeing far more orders to buy the shares in question compared with the number of orders to sell them, thereby concluding that their price is going to rise.

The same spoofer might then buy the shares themselves, causing the price to rise. When it does so, the spoofer would cancel its buy orders in order to profit from the owned shares it sells.

Although it isn't at all clear that automated trading is any more dangerous than the human trading it is replacing, we are still left with a sense at the end that we have entered an era of machines—and those who own them—are controlling everything.

Who Rules the Technology?

In 2000, Harvard Law School Professor Lawrence Lessig predicted that the Internet would become an apparatus that tracks our every move, erasing important aspects of privacy and free speech in our social and political lives. "Left to itself, he said, "cyberspace will become a perfect tool of control." [122]

That surveillance follows us everywhere and from everywhere.

More than 1,700 satellites monitor our planet. From a distance of about 300 miles some can discern a herd of buffalo or stages of a forest fire.

Our skies have become cluttered with drones. About 2.5 million were reportedly purchased by American hobbyists and businesses in 2016. This doesn't include a huge fleet of unmanned aerial vehicles used by the U.S. government against terrorists and illegal immigrants.

Cities are rapidly expanding CCTV networks in the interest of public security. New York City ramped up installations following the September 2001 attacks to roughly 20,000 officially-run cameras by 2018 in Manhattan alone. In 2018, Chicago reportedly had a network of 32,000 CCTVs to help combat the inner city murder epidemic. Thanks to federal grants, Houston, which as recently as 2005 had none, had about 900 by 2018, with access to an additional 400.

George Orwell's grim 1984 story admonition that "Big Brother is watching you" has a rapidly growing number of progeny. That same year he wrote it—in 1949—an American company released the first commercially available CCTV system. Two years later, in 1951, Kodak introduced its Brownie portable movie camera.

An estimated 106 million new surveillance cameras are now currently being sold annually in the United States. Tens of thousands of cameras known as automatic number plate recognition devices, or ANPRs, hover over roadways to catch speeding motorists and parking violators.[123]

Cameras connected with facial recognition are being used to by the Transportation Safety Administration (TSA) air marshals to trail and closely monitor unsuspecting Americans targeted for special airport inflight surveillance. As reported by the *Boston Globe*, TSA's "Quiet Skies" program, which began in

2010, entitles and enables teams of undercover agents to document whether targeted individuals…

> …change clothes or shave while traveling, abruptly change direction while moving through the airport, sweat, tremble or blink rapidly during the flight, use their phones, talk to other travelers or use the bathroom, among many other behaviors.

The targets are not necessarily people who have done anything that warrants any previous reasons to be on a terrorist watch list, although it is intended to identify those who "flag" reasons for concern. The first red flag is foreign travel—specifically, frequent visits to "countries that we know have a high incidence of adversarial actions."

Risk assessment targeting is far from reliable. The Globe reports that a flight attendant and a federal law enforcement officer are among those who have been flagged for surveillance under the program.[124]

There are understandably valid reasons for citizens to accept reasonable tradeoffs between gaining more security at the cost of less privacy in public venues. Many of us will recall footage from security cameras that cracked cases of the 2005 London subway and 2013 Boston Marathon bombings…and when Eric Cain was caught on camera shooting a Tulane University medical student named Peter Gold in 2015 (after Gold prevented him from abducting a woman on the streets of New Orleans).

And it's beyond question that our Western nation attitudes have changed over the past century.

William Webster, a professor of public policy at the University of Stirling in Scotland, notes that the pre-World War II rhetoric about public safety was:

'If you've got nothing to hide, you've got nothing to fear.' In hindsight, you can trace that slogan back to Nazi Germany. But the phrase was commonly used, and it crushed any sentiment against CCTVs.[125]

Former U.K. deputy prime minister, Nick Clegg has observed:

And basically, it's happened without any meaningful public or political debate whatsoever. Partly because we don't have the history of fascism and nondemocratic regimes, which in other countries have instilled profound suspicion of the state. Here it feels benign. And we know from history, it's benign until it isn't.[126]

Carnegie Mellon University professor of information technology Alessandro Acquisti has us remember:

In the cat-and-mouse game of privacy protection, the data subject is always the weaker side of the game.

Acquisti reminds us that we in America haven't been through the experience of the man in the brown leather trench coat knocking on the door at four in the morning...so when we talk about government surveillance, the resonance is different. He warns:

[The desire for privacy] is a universal trait among humans, across cultures and across time. You find evidence of it in ancient Rome, ancient Greece, in the Bible, in the Quran. What's worrisome is that if all of us at an individual level suffer from the loss of privacy, society as a whole may realize its value only after we've lost it for good.[127]

As Gus Hosein, executive director of Privacy International, notes:

> [If] the police wanted to know what was in your head in the 1800s, they would have to torture you. Now they can just find it out from your devices.[128]

Personal monitoring devices are proliferating...dash cams, cyclist helmet devices to record collisions, doorbells equipped with lenses to catch package thieves and inexpensive sound and movement-activated home security cameras are becoming ubiquitous.

As Rachel Holmes accurately observes in her Guardian.com article, it's now clear that the digital world has evolved into a creature of control. We realize that we are being watched, monitored and recorded and don't seem to care.

Facebook and other corporate Internet giants gather data to maximize profits from our consumer habits, from grocery shopping to TV viewing patterns to political interests and affiliations. Holmes writes:

> Just like trawlers with dragnets, all sorts of other collateral data gets hauled in along the way. Data surveillance, once intangible and invisible, now blatantly announces its presence in our everyday lives. Mobile accessories and interconnectivity between gadgets and appliances in our homes—the Internet of Things—create an unprecedented network of tracking devices capturing data for commerce and government.[129]

Holmes points out that the technology we thought we were using to make life more efficient started using us some time ago, and it's now attempting to reshape our social behaviors into patterns reminiscent of the total surveillance culture. She writes:

> In an increasingly online everyday life, our use of social media has become a medium for normalizing the acceptability of intrusion and behavioral connection. We are bombarded by 'helpful recommendations' on education, health, relationships, taxes and leisure matched to our tracked user profiles that nudge us towards products and services to make us better citizen consumers. The app told us that you only took 100 steps today. The ad for running shoes will arrive tomorrow.

U.S. lawmakers have expressed concerns about tech giants including Google, Facebook, and Twitter on a range of issues such as privacy and manipulation of content by foreign actors.

Google's vice president for public policy Susan Molinari admitted in a letter to senators that the company allows app developers to scan Gmail accounts, although in 2017 the company stopped its previous practice of ad targeting. Using software tools provided by Gmail and other email services, outside app developers can still access information about what products people buy, where they travel and which friends and colleagues they interact with most.

In some cases, employees at these app companies have read people's emails to improve their software algorithms. Google policies also allow their own employees to read emails so long as that practice is relevant to the functioning of their apps and disclosed to users.[130]

The app developers generally are free to share the data with others if their privacy policies adequately disclose possible uses. Molinari's letter says:

> *Developers may share data with third parties so long as they are transparent with the users about how they are using the data.*

A bipartisan group of senators wrote to Commerce Secretary Wilbur Ross saying that recent revelations about the misuse of data "demonstrate how little we know about who has access to consumers' private information and how the data is used."[131]

Top tech executives, Facebook COO Sheryl Sandberg and Twitter CEO Jack Dorsey, were also grilled in a September 6, 2018 Senate open hearing regarding their companies' roles in intentionally stifling conservative voices on their social media platforms. Alphabet Inc.'s Google was invited to testify, but declined, leaving a third adjacent seat conspicuously empty.[132]

This hearing topic should be recognized as much larger and fundamentally more important than which direction any ideological bias favors—whether to the political left or right. These companies, which hold monopolistic control over social media and search platforms, wield tremendous influence over our open access to ALL public and private information and opinion discussions.

Their status as corporate—rather than government—entities, entitles them to determine what we are allowed to see, what we are not allowed to see, and from whom, entirely at their own discretion. Moreover, they can do so without having to justify their specific-case decisions or explain their rationale.

In 2012, Federal Trade Commission staff issued a report recommending a lawsuit against Google—which controls 90 percent of all Internet searches—for anti-competitive conduct. The commission, led by Obama appointee Jon Leibowitz, subsequently voted against it.

Together, Facebook and Google, which control 60 percent of all digital ad revenues, have used their market dominance to undercut competitors. Facebook has arbitrarily blocked publishers from promoting factual and opinion content on its news stream by tagging certain ads as "political."

Facebook CEO Mark Zuckerberg said the undisclosed algorithm changes would focus on news from "trustworthy" publications that people find "informative."

Responding to accusations that Twitter discriminates against conservatives, Jack Dorsey admitted during an August 18, 2018 *CNN* interview:

> We need to constantly show that we are not adding our own bias, which I fully admit is left-leaning. But the real question behind the question is, are we doing something according to political ideology or viewpoints? And we are not. Period.[133]

Nevertheless, this doesn't explain why numerous scholarly and informative PragerU videos posted on Twitter and Google's YouTube were removed as "hate speech." The rational for "hate speech," as Facebook puts it, is "dehumanizing language." Yet none of the 50 or so censored videos—with as many as 3 million followers—included violent, sexually explicit or hateful content that violated Google's stated Community Guidelines.

Google's transparently left-leaning tilt is impossible to ignore. The company went all-in for Hillary during the 2016 presidential elections. Company employees donated $1.6 million to her campaign, and Eric Schmidt, Google's Executive Chairman from 2001 to 2015, provided direct political data analytics support.

The *Wall Street Journal* reported that just days after the Trump administration instituted a January 2017 travel ban, Google employees discussed ways they might be able to tweak the company's search-related functions to show users how to contribute to pro-immigration organizations and contact lawmakers and government agencies.[134]

The original travel ban was implemented to restrict immigration from countries deemed a security risk. It temporarily barred visitors and immigrants from seven Muslim-majority countries, and placed new limits on the U.S. refugee program.

Google joined nearly 100 technology companies, including Apple Inc. and Facebook Inc., in filing a February 2017 joint amicus brief challenging the travel ban. An email from an employee of the Search Product Marketing Division explained that there was a "large brainstorm" going on throughout the company's marketing division. It stated:

> *Overall idea: Leverage search to highlight important organizations to donate to, current news, etc., to keep people abreast of how they can help as well as the resources available for immigrations [sic] or people traveling.*

Another Google employee email said:

> *I know this would require a full on sprint to make happen, but I think this is the sort of super timely and imperative information that we need to know that this country and Google would not exist without immigration.*[135]

Google is known to have been developing a "Dragonfly" search engine for China that blocks queries for terms including freedom of speech, religion and democracy. The algorithm is also reportedly designed to link users' searches to their phone numbers. That, in turn, could make it easier for Chinese authorities to identify people searching for banned topics.

In response to scrutiny and criticism, a Google spokesperson said of Dragonfly:

> We've been investing for many years to help Chinese users, from developing Android, through mobile apps such as Google Translate and Files Go, and our developer tools. But our work on search has been exploratory, and we are not close to launching a search product in China.

Google's China-specific search engine, Google.cn, launched in 2006 as a means for the company to stay in the country while abiding by its strict censorship rules. The company's normal search engine was still technically available, but was heavily filtered by China's "Great Firewall," a system of blocks that lets the government control the web content Chinese citizens can view.

Critics charge that China could use the new Dragonfly to replace air pollution information in online news reports, giving the appearance that pollution levels aren't as dangerous as they actually are—something that China has already been accused of.

Google eventually bailed on the project in 2010 amid China's accusations that the company allowed pornography on its search engine. A sophisticated hacking attempt on Google that originated in China was the last straw, at least temporarily prompting and the company to stop censoring content there and move its operations from the mainland.

While Google says it's only exploring a new search engine for the Chinese market, the tech giant already has an existing presence in the country. China is the world's most populous country with a growing middle class which is becoming increasingly connected. Abandoning China would cede the country's huge market to homegrown competitors like Baidu, China's largest search engine.

Nevertheless, U.S. lawmakers are understandably lambasting Google for even considering working with the Chinese government again. Senator Tom Cotton (R-AR) issued a statement in August 2018, calling out the tech giant for potentially working on a new Chinese information censorship search engine. He wrote:

> *Google claims to value freedom and one hopes Google will put its corporate principles and America first, ahead of Chinese cash.*[136]

Enormous and ever-expanding national and global power and influence wielded by information tech giants warrant judicious contemplation. As Rachael Holmes urges us to recognize:

> *The fact that billions of us now use a handful of corporate-owned global platforms to manage pretty much every aspect of our daily lives indicates how fast the potential of digital culture is shrinking. A tectonic rift exists between the corporate apparatuses of Google, Facebook, WhatsApp (now owned by Facebook), Twitter, Apple and Amazon, and the open source and creative commons ideals of the Internet of public good.*[137]

Holmes warns that we dangerously risk allowing ourselves to become a vast network of informants on each other and ourselves, a wireless connected system where GPS-based location tracks us on our mobile phones; one where social media apps monitor, record and pass along personal information about our spontaneous thoughts, social lives and connected relationships; one which has no national borders; one which does not recognize outmoded distinctions between state and corporate power, citizens and consumers and platforms and products.

We have become a society of individuals who behave on a spectrum of willing complicity with the demise of our privacy.

Rachael Holmes writes:

> *We know Twitter is broadcasting. We're generally naïve about limits to our visibility and history on Facebook. We're in self-denial where there ought to be reasonable expectation of privacy, as with texts, emails and online shopping. In truth we already know nothing's private because the corporates and government can get it anyway. And the icing on the cake is that we are paying for our surveillance out of our own hard-pressed pockets.*[138]

So, after all, who rules the technology?

It is those who develop the algorithms; and those who control the secured gateway on ramps, toll booths and draw bridges along the Internet super highway.

The power resides with purse-string politicos and big tech companies who hold sway over smart cities with smart cars wirelessly connected to a vast Internet of Things network which extends from our home appliances—to our places of work—to our sidewalks and streets—to electronic eyes and ears that track and display our every movement on remote facial recognition monitors.

The rulers are those who control enormously prosperous, unaccountable and ideologically and politically biased digital access and social media companies that manipulate and censor information to fit their preferred narratives.

In summary, the technology rulers are all those whom the rest of us relinquish our long-cherished rights of privacy and self-determination to. There are many of them. They are not only winning…we are generously paying them to do so.

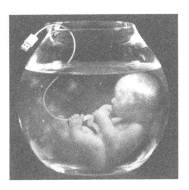

A New Evolutionary Era?: Influences Upon Culture and Values

GREEK/FRENCH PHILOSOPHER Cornelius Castoriadis argued that individuals in most societies do not depend on themselves to lay down their own laws through "autonomy," but rather assume that law is created by some external force found beyond themselves, whether by gods, nature, history or reason ("heteronomy"). In today's world, computer algorithms are gaining ever-greater heteronomous influences over our social, economic and broader cultural lives and values.

In his book *The Master Algorithm: How the Quest for the Ultimate Learning Machine Will Remake Our World,* University of Washington professor of computer science and engineering professor Pedro Domingos describes how self-learning machines are changing our everyday lives—from social networks and science to business and politics right up to the way modern wars are waged. The book, which drew praise from Microsoft founder Bill Gates and Google CEO Eric Schmidt, characterizes five rival technical and social orientations regarding primary algorithm development priorities and outcomes:

- Symbolists view computer learning as the inverse of

deduction and take ideas from philosophy, psychology and logic.

- Connectionists aspire to reverse engineer the brain and are inspired by neuroscience and physics.

- Evolutionaries simulate evolution on the computer and draw upon genetics and evolutionary biology.

- Bayesians believe that learning is a form of probabilistic inference and have their roots in statistics.

- Analogisers learn by extrapolating from similarity judgements and are influenced by psychology and mathematical optimization.[139]

Domingos discusses his quest to find the "Master Algorithm," one that combines key features of all five. If found, he believes, "the Master Algorithm can derive all knowledge in the world—past, present, and future—from data."

As Domingo told the German magazine Der Spiegel, he found it "both exciting and scary" to learn that a copy of his book was seen on Chinese President Xi Jinping's bookshelf. It was exciting because China is developing rapidly, and there are all sorts of ways the Chinese and the rest of the world can benefit from AI. It is scary because this is an authoritarian government going full tilt on using AI to control their population. He said:

In fact, what we are seeing now is just the beginning. Like any technology, AI gives you the power to do good and evil. So far, we have been focusing on the power to do good, and I think it is enormous. But the power to do evil is there, too.[140]

Information Technology Masters and Martinets

As discussed in the Stanford Encyclopedia of Philosophy:

> *Everything we do and everywhere we go generates information: all of your economic transactions, a GPS generated plot of where you traveled, a list of all the web addresses you visited and the details of each search you initiated online, a listing of all your vital signs such as blood pressure and heart rate, all of your dietary intakes for the day, and many other examples can be imagined.*[141]

This pervasive intrusion of information technology into our lives and society has profound moral implications. As this data gathering becomes increasingly automated and ever-present, we must ask a variety of important questions. For example: who is in control of this data; what is to be done with it; who will insure its accuracy; which bits of information should be made public or held private; and which should be allowed to become the property of third parties, such as corporations?

The control of information empowers big data organizations to wield great ethical influences over virtually all types of traditional social and economic institutions including religious organizations, universities, libraries, healthcare officials, government agencies, banks and corporations.

The Stanford analysis points out that while some may argue that this situation is no different from the moral issues revolving around the production, access and control of any basic necessity of life, there is an important basic difference.

In traditional moral issues, if one party controls the access of some natural resource, then that, by necessity, excludes others from using it. This is not necessarily so with digital information because it is non-exclusionary in the sense that we can all at least theoretically possess the same digital information. Copying and exchanging it from one digital source to another does not require eliminating the previous copy.

The Stanford report explains:

> Modern information technology makes recording of information easy, and in some cases it is done automatically. A growing number of people enter biometric data in order to live healthier lifestyle choices; tracking websites can make website pages load faster the next time you visit them...yet if someone were following you around a library noting down this kind of information you might find it uncomfortable or hostile...online this kind of behavior takes place behind the scenes and is barely noticed.

According to some professionals, information technology has all but eliminated the private sphere. Scott McNealy of Sun Microsystems famously announced in 1999: "You have zero privacy anyway. Get Over it." [142]

In her paper *Toward an Approach to Privacy in Public: Challenges of Information Technology,* Hellen Nissenbaum writes:

> [W]here previously, physical barriers and inconvenience might have discouraged all but the most tenacious from ferreting out information, technology makes it available at the click of a button or for a few dollars. [143]

Nissenbaum points out that information technologies that store user data in the cloud—on a device remote from ownership and operation by users—convenience is traded off for loss of control. While the user can access data from any device she or he happens to be using from anywhere, so can others. Such ease of access makes one's relationship with their own data more tenuous, with added uncertainty about the physical location of that data.

Increasingly, machines make important life-changing financial decisions about people without much oversight from human agents. Whether or not you will be given a credit card, receive a mortgage loan, and the price you will have to pay for insurance, for example, are very often determined by a machine.

This being the case, Hellen Nissenbaum asks:

> *Who has the final say whether or not some information about a user is communicated or not? Who is allowed to sell your medical records, your financial records, your friend list, your browser history, etc.?* [144]

In a very bad-case scenario, consider China and its collaboration with Zimbabwe and other African countries as a draconian surveillance state model.

China is building the world's largest facial recognition database in support of mass surveillance and incarceration programs. Special emphasis is directed to the western province of Xinjiang, where the population is predominantly Muslim Uighurs of Turkish descent.

In 2015, the Chinese government launched its "Made in China 2025" policy to dominate cutting-edge technologies industries. This was followed in 2017 by plans for the country to be a world leader in the field of artificial intelligence by 2030 and to build a $250 billion industry.

Although the developing world is a big part of these ambitions, China doesn't just want to dominate these markets. It is also using developing countries as laboratories to improve its surveillance technologies.[145]

Many parts of Africa are already essentially reliant on Chinese companies for their telecoms and digital services. Transsion Holdings, a Shenzhen-based company, was Africa's number one smartphone company in 2017.

CloudWalk Technology, a Guangzhou-based start-up, has signed a deal with the Zimbabwean government to provide a mass facial recognition system. As planned, it will enable this country with a bleak human rights record to improve its facial recognition system's accuracy for dark-skinned populations. This technology limitation has caused previous misidentification problems.

During 2017, nearly half of global investment in AI went to Chinese start-ups. CloudWalk alone received a $301 million grant from the Guangzhou municipal government.

Maya Wang, a senior China researcher for Human Rights Watch, said:

> We are concerned about the deal, given how CloudWalk provides facial recognition technologies to the Chinese police. We have previously documented [the Chinese] Ministry of Security's use of AI-enabled technologies for mass surveillance that targets particular social groups, such as ethnic minorities and those who pose political threats to the government.

Some Zimbabweans are concerned about how their data will be used. An unnamed Ph.D. student at Beijing Normal University said:

> [T]he question is what the Chinese company will do with our identities…It sounds like a spy game." He added that he "know[s] for a fact that the Zimbabwe

government will use this tech to try and control people's freedom.[146]

As reported in *Foreign Policy*, communications are routinely monitored and censored in China. Poorly chosen words sent on WeChat, the main messaging platform, can land the sender in jail.

This principle of intrusive surveillance is also well established in Zimbabwe. A 2017 Cybercrime and Cybersecurity Bill criminalized communicating falsehoods online, just as restricting legal rhetoric has been used to silence dissent in China. Now, with the technology provided by the CloudWalk deal, government opponents will have even fewer places to hide.

Another African country following the Chinese model is Tanzania, where the vague idea of "content that causes annoyance" is now verboten. China and Tanzania have both banned posting of "false content" and are cracking down on expunging "decadent" material from social media networks.

Such governments justify censorship and retribution as a security excuse to preserve law and order. In this regard, China is not the only country to market advanced surveillance technology to abusive governments. Between 2014 and 2017, nearly one-third of licenses were granted by the European Union to export surveillance products to countries deemed "not free" by the watchdog Freedom House.

Trading on Good and Bad Faith

As discussed in the previous chapter, computer algorithms have come to exert huge impacts upon our U.S. business economy, stock movements and valuations in particular. Wall Street quantitative analysts (Quants) and masters of "algo trading" devise and apply "evolutionary algorithms" which allow them to create vast empires of wealth through manipulations of complex stock portfolio credit structures called derivatives.

Financial experts and wonkish talking heads earn big incomes endeavoring to explain these mysterious financial instruments to lay audiences. Warren Buffett, the Wizard of Omaha, once characterized derivatives "weapons of financial mass destruction." [147]

Back in 2007, some of those genius quants—the best and brightest geeks Wall Street firms could buy—fed $1 trillion in subprime mortgage debt into their supercomputers, added some derivatives, messaged the arrangements with computer algorithms, and—poof!—created $62 trillion in imaginary wealth. That wealth soon also poofed away. On September 29, 2008, the Dow Jones Industrial Average fell 777.68 points in intra-day trading.

Richard Dooling observed at that time in the *New York Times* that as the financial crisis spread (like a computer virus) on the Earth's nervous system (the Internet), it was worth asking if we have somehow managed to colossally outsmart ourselves using computers. He wrote:

> *It's not much of a stretch to imagine that all imaginary wealth is locked up somewhere inside the computers, and that we humans, led by the silverback males of the financial world, Ben Bernanke and Henry Paulson, are frantically beseeching the monolith for answers. Or maybe we are lost in space, with Dave the astronaut pleading, "Open the bank vault doors, Hal.*[148]

After all, stock transactions derived from much simpler beginnings. The term "stock" originally referred to a stick of hazel, willow or alder wood, inscribed with notches indicating monetary amounts and dates. When funds were transferred, the stick was split into identical halves. One went to the depositor, the other to the party safeguarding the money who represented positive proof that the agreed amount value was backed up by gold.

That arrangement was good enough to last over 600 years, until later generations decided that more speed and efficiency was needed.

George Dyson, historian of technology and author of *Darwin Among Machines,* observed that making money has become all about the velocity of moving it around so that it can exist in Hong Kong one moment and Wall Street a split second later. He wrote:

> *The problem starts as the [then] current crisis demonstrates, when unregulated replication is applied to money itself. Highly complex computer-generated financial instruments (known as derivatives) are being produced, not from natural factors of production or other goods, but purely from other financial instruments.*[149]

Richard Dooling added:

> *It was easy enough for us humans to understand a stick or a dollar bill when it was backed by something tangible somewhere, but only computers can understand and derive a correlation structure from observed collateralized debt obligation tranche spreads. Which leads us to the next question: Just how much of the world's financial stability now lies in the 'hands' of computerized trading algorithms?*

As Yuval Noah Harari writes in *Sapiens: A Brief History of Humankind,* the "most important economic resource is trust in the future." And key to that trust is confidence in a universal medium of exchange—money—which enables people to convert almost everything into almost everything. Ideal forms of money also provide a means to store wealth and to transport it from place to place.

Harari logically notes here that some forms of wealth, such as real estate, cannot be transported at all. Commodities such as wheat and rice can be transported only with difficulty. He writes:

> *Imagine a wealthy farmer living in a moneyless land who emigrated to a distant province. His wealth consists mainly of his house and rice paddies. The farmer cannot take with him the house or paddies. He might exchange them for tons of rice, but it would be very burdensome and expensive to transport all that rice.*[150]

Money's ability to convert, store and transport wealth easily and cheaply gave rise to complex commercial networks and dynamic trading markets. Cowry shells served this purpose for about 4,000 years all over Africa, South Asia, East Asia and Oceania. Yet like dollars today, their value existed only in common and imagined trust. Determinations of their worth are entirely independent of any inherent value attached to the chemical structure of the shells and paper, or in their colors or their shapes.

China had developed a monetary system during the seventh century BC which was based upon bronze coins and unmarked silver and gold ingots. The entire world later relied heavily upon gold and silver along with a few trusted currencies such as the British pound and the American dollar.

Currency trading emerged as merchants traveling between India and the Mediterranean noticed differences in the value attached to gold. In order to make a profit, they would buy gold cheaply on India and sell it dearly in the Mediterranean. This caused the demand and value of gold in India to skyrocket, while at the same time, the Mediterranean would experience an influx of gold, with its value consequentially dropping. Within a short time the value of gold in India and the Mediterranean would be quite similar.

Harari asks, (and answers):

> Why are you willing to flip hamburgers, sell health insurance or babysit three obnoxious brats when all you get for your exertions is a few pieces of colored paper? People are willing to do such things when they trust the figments of their collective imagination.

Trust is the raw material from which all types of money are minted. It follows, that while money is accordingly a system of mutual trust, it is not "just any system of mutual trust. Money is the most universal and most efficient system of mutual trust ever devised." [151]

Harari observes that trust in this context involved a very complex and long-term network of political, social and economic relations. He reflects:

> Why do I believe in the cowry shell or gold coin or dollar bill? Because my neighbors believe in them. And my neighbors believe in them because I believe in them. And we all believe in them because our king believes in them and demands them in taxes, and because our priest believes in them and demands them in tithes.

The crucial role of trust, he explains, is why our financial systems are so tightly bound up with our political, social and ideological systems, why financial crises are often triggered by political developments, and why the stock market can rise and fall depending on the way traders feel on a particular morning.

The advent of AI information technologies and the Internet have now introduced a purportedly new cryptocurrency blockchain trading revolution which demands still a greater leap of public faith. A blockchain is a decentralized, distributed ledger used to record financial transactions across many computers.

This is accomplished in a way that the record cannot be altered retroactively without alteration of all of the subsequent "blocks" and the consensus over the entire network. This allows individual participants to continuously and inexpensively verify and audit their transactions over an autonomously managed peer-to-peer network system incorporating a distributed timestamping server.

George Glider, author of *Life After Google: The Fall of Big Data and the Rise of the Blockchain Economy,* describes blockchain and cryptocurrency as "a new technological revolution that is rising as we speak." In a *Wall Street Journal* interview with Stanford University's Hoover Institution fellow Tunku Varadarajan, Varadarajan says that blockchain has already generated "a huge efflorescence of peer-to-peer technology and creativity, and new companies." [152]

Glider argues that the decline of initial public offerings in the United States has already been redressed by the rise of the ICO, the "initial coin offering," which in 2017 raised some $12 billion for several thousand companies. He opined that with the cryptographic revolution:

> We're now in charge of our own information. For the first time in history, really, you don't have to prove who you are, or what you are, before a transaction.

This is because a blockchain allows users "to be anonymous if they wish, while also letting them keep a time-stamped record of all their previous transactions. It allows us to establish unimpeachable facts on the Internet." This, he proposes, evokes trust in the Internet, "without having to trust or rely on [Google co-founders Sergey Brin and Larry Page], [Facebook co-founder] Mark Zuckerberg, or whoever the paladins of the new economy may be."

Mister Gilder asserts that this development constitutes a notable victory for mankind.

Well, perhaps. But there are a couple of bugs that might need to be worked out before mankind takes that victory lap.

For many, the cryptocurrencies market holds great appeal as a purported get-rich-quick opportunity. Marketing hype abounds offering wild speculations such as "BitCoin going to hit $1 million!"…Ripple going to replace fiat (country-based) currency!"

Fundamentally, the actual exchange of cryptos and the value placed on them is intrinsically based on 1's and 0's, the same makeup of all things digital. And like everything traded, high demand drives up cryptocurrency price, while conversely, low demand causes the assumed value to drop.

So far, the same circumstance is true regarding the stock market, stock options and other currencies. Supply and demand principles apply to them as well. The key difference with cryptocurrencies is that these other investments typically also have "assets" backing their valuations.

For example, when you invest to purchase stock in a company, the incentive is to believe there is a good chance that it will continue to create good products and services, will continue to grow and will pay reasonably profitable and stable return-on-investment dividends. If they fail, your initial investment will go down, but you still retain some percentage of their assets, their intellectual prowess and their future business prospects.

This same logic applies to purchasing a house. Here again, your assessment of current and future supply and demand will likely affect what you are willing to invest. But that property investment, which includes money spent for the house, together with the land it is on, will always retain at least some worst-case asset value.

With cryptocurrencies, those 1's and 0's aren't backed by any assets at all—their only value is based upon an expectation that someone else will be willing to pay more than you did, recognizing that you will lose value when and if they don't.

Fiat currencies (those issued by governments) are very easy to spend and are a globally recognized form of payment. You can take a Baht, or a Canadian dollar, or a U.S. dollar to a bank and get it exchanged to any currency you prefer. Granted, they are often backed by promises made by some pretty shaky country economies, but then, you naturally take this liability into consideration.

Wherever you go in the world, even the concept of money traditionally backed by trust in governments is being challenged within a new cyber domain to stretch our conventional ideas regarding what "trust in money" really means.

Wherever government control of money supply chains exist there are likely to be many legitimate things to be suspicious and to worry about. For example, the country—any country—can simply print more money which to be distributed within banking networks that proficiently game the system. Nevertheless, you can still always spend that currency.

In reality, cryptocurrencies aren't true currencies at all, but are instruments that can be traded. In order to get your money out and in order to actually spend it, a newcomer on the other end is required to pay for it. Although there is a limited supply of most cryptos, meaning that this can inflate the demand for them, they aren't backed by anything with agreed value: no gold, silver or bronze transacted in the form of cowry shells; no grain holdings recorded on hazel, willow or alder wood stick inscriptions.

In short, crypto currencies bear an unmistakably close resemblance to Ponzi schemes.

Yuval Noah Harari's characterization of money as "the most universal and efficient system of mutual trust ever devised" raises still another important question regarding such trust applied to crypto currencies, namely a trust that they can be made secure from crypto criminals.

Although many will argue that blockchain technology is very secure, at least presently, cryptocurrencies are vulnerable to third-party invasions. Included are cryptocurrency exchanges, hardware creators, crypto miners and lots of other shady folks and entities in the Internet-connected cryptoworld.[153]

A 2013 hack of Mt. Gox, a Shibuya, Tokyo-based exchange which then controlled 70% of the world's bitcoin trading, was discovered in 2014 following transgressions which dated back to 2011. The intruders hacked into a Mt. Gox auditor's computer, transferred a large sum of Bitcoins to themselves, and then used the exchange's software to sell all the Bitcoin, resulting in a sudden drop in Bitcoin value. Although the price was later re-adjusted, the hacker made off with about $8,750,000.

Mt. Gox did not have any version control software, meaning that any coder could accidentally overwrite their colleague's code if they were working on the same file. With a lack of coding security, hackers were able to tamper with transactions made on the exchange. The company went bankrupt, and is no longer in operation.

In August 2016, Bitfinex, one of the world's largest crypto exchanges, announced that nearly 120,000 Bitcoins had been purloined from its users' accounts. The strangest aspect of this hack was that the drained Bitcoins came from multi-signature accounts which are considered to be one of the safest digital signature schemes in the industry. "Multi-sig" transactions can only be authorized with the presence of signatures of multiple parties who manage the funds to eliminate risks of having them stolen.

The transfer of funds on a multi-sig account requires access to all of these keys (usually 3), for any transaction. In this instance, Bitfinex had held two of these keys, while its partner, BitGo, a blockchain security company that had formed its multi-sig system, had access to the third. Yet, in spite of this tight security measure, the hackers were able to gain access to all three of the keys to siphon off users' Bitcoins to an unknown address. In the aftermath of the hack, neither Bitfinex nor BitGo admitted to any wrongdoing.

In December 2017, NiceHash, a Slovenian crypto mining service, was at the center of another large-scale hack. The service allowed the owners of cryptocurrency mining equipment to rent out their hash power to customers over short time spans. On December 6, 2016, Marko Kobal, the CEO of NiceHash, reported on *Facebook Live* that attackers of the service had siphoned away more than $80 million.

Few details were revealed other than that the heist transpired from an employee's compromised computer, through which the hackers were able to use that employee's credentials to access to the company's NiceHash system. The Bitcoins were sent to an unknown address, one that neither NiceHash nor the rightful owners could access. NiceHash briefly discontinued operations for 24 hours for a checkup and analysis of the hack, then recommended that users change their passwords for additional security. The CEO resigned from the company after the hack.

In still another cryptocurrency hack, cybercriminal thieves hit Coincheck, a Japanese cryptocurrency exchange, in January 2018 for a whopping $534 million in NEM coins stolen from a "hot wallet" connected to the Internet. Similarly to the Bitfinex hack, many of these altcoins were held in multi-sig accounts.

Following the hack, Coincheck released a statement in which it declared its opposition in keeping cryptocurrency in hot wallets, and announced that it intends on recompensing those who lost crypto in the hack. Users who had their NEM stolen will receive $0.83 per NEM. A complete refund will cost Coincheck about $420 million.

A New Normal: Living With Cybersecurity Threats

In September 2018, Facebook reported that hackers had gained access to nearly 50 million accounts in the largest-ever security breach at the social network. Executives said that the attack was sophisticated, requiring the hackers to find and exploit three obscure flaws in the code. They had succeeded in exploiting three bugs in the "view as" feature which let the outsiders steal access tokens—digital keys that keep people logged into Facebook.[154]

With the stolen tokens in hand, hackers can then take over accounts, impersonate users and access private information about them and their friends, including a user's Facebook connections, friends' posts and messages.

Such breaches can give hackers access to information that could be used to pirate credit card accounts and commit identity theft. The culprits can then either use such information themselves, or sell it within a global market space to other criminals.

The Internet of Things has opened up a superhighway for fraud, theft and other cybercrimes to painfully erode citizen confidence in the security of their work and private lives.

Like cancer cells, Distributed Denial of Service (DDoS), phishing and ransomware attacks take advantage of the naturally occurring resources of the host organism. While our bodies have immune cells that are deployed to fight tumor formation, we as a society have yet to develop the requisite immunity against cyberattacks within a shared global communications system. Moreover, humankind is inadvertently inventing new cyberweapons which can be targeted against themselves at a blinding speed.

Writing in the *Wall Street Journal*, Ralph Nader discussed cybersecurity threats to driverless cars from hacking—the same sorts of malware cyberattacks and data breaches that have occurred in other sectors of the economy. He warns that without critically needed protections, hackers could seize control of a driverless car. A best-case scenario would result in an annoying glitch in which the infotainment screen freezes.

In the worst case scenario, a vehicle, or an entire model line or fleet, could be hijacked into an autonomous weapon, leading to highway mayhem.[155]

When malware and spyware is created by state actors, we enter the world of informational warfare. Every developed country in the world experiences daily cyberattacks, with the major target being the United States.

Cyberwarfare is broadly regarded to represent the most complex national security threat the United States has ever faced, most particularly because the technology is changing so rapidly.

A 2012 report by the National Research Council concluded that a cyberattack on the U.S. energy grid could black out a large region of the nation for weeks or even months. The consequences of such an extended power outage are beyond comprehension. Deaths resulting from losses of vital life support systems in hospitals, disrupted electricity for food preservation, clean water supplies and sanitation and a massive breakdown of the transportation system are but a few examples of carnage.[156]

A limited preview of what might be experienced as a grid cyberattack resulted when a 2003 extreme weather blackout left

50 million people in the Northeastern United States without power for four days. Economic losses alone were variously estimated to be between $4 billion and $10 billion.

Russia, Iran, North Korea and others are known to have large-scale, offensive cyberattack programs. According to the Department of Homeland Security and the FBI, Russia appears to be laying a foundation for a large scale cyberattack on U.S. infrastructure. "Dragonfly 2.0" hackers identified by DHS as Russian cyber actors engaged in an extensive plan to disrupt U.S. power plants and computer networks controlling the grid.[157]

Governments, including that of the United States, are building "cyber armies" to take on attackers. Such offensive and preemptive risk mitigation cyber policies are shaped by trade-offs between deterrence on the one hand and intelligence collection and diplomatic standing on the other. A relaxed cyber engagement policy increases U.S. deterrence capabilities—if you hit us, we can hit back. But it could also endanger existing spy operations.[158]

Such conflicts are not new or unique to cyberspace. What the Pentagon calls "intelligence gain/loss" considerations are applicable to all domains. Dropping a bomb on a terrorist camp may disrupt one plot, but it may also kill the terrorist group's courier who is under surveillance.

Yet, as Dave Weinstein points out in a *Wall Street Journal* article:

> [I]n the digital domain, these calculations become much more complicated and unpredictable. Unlike the physical realm, where it is easy to calculate the blast radius of an ordinance or the likelihood of civilian casualties, the collateral effects of a cyber operation are often best guesses.

Those "best guesses" can rapidly spin out of control. A 2017 Russian cyberattack that became known as Not-Petya began as a targeted operation against organizations in Ukraine and metastasized into a global campaign that struck some of the world's largest corporations. Soft targets included the American drug manufacturer Merck, the Danish shipping giant Maersk, and even the Russian state-owned oil company Rosnett.

Ironically, that cyber weapon was allegedly developed by and later stolen from the U.S. National Security Administration. It then travelled well beyond its original targets.

In any case, despite the challenges and risks of operating militarily in this new domain, allowing a defensive status quo to prevail is an indefensible national security option.[159]

Information security threats are now rocketed up as the result of staggeringly large processing capacities afforded by rapidly accelerating quantum computer (QC) advancements. Just as AI promises to transform an endless variety of peaceful information and problem-solving tasks, its vast capacity to out-think conventional computers presents enormously troubling cybersecurity challenges.

In addition to overwhelming cryptographic codes used to protect top secret data, it can also be weaponized for conduct of armed conflicts at much larger scales and higher speeds than humans can comprehend or react to.

International foes and friends are racing to achieve QC supremacy which can defeat all current-generation defenses against military, information security, banking and utility infrastructure system cyberattacks. The first hostile nation to win this race will be able to open the encrypted secrets of every country, company and person on the planet; dominate global information-technology and the global financial systems; compromise the safety of medical, food and water services; put transportation and energy infrastructures at risk; and threaten domestic and military security systems.

Researchers Daniel Bernstein at the Technische Universiteit Eindhoven and Tanja Lange at the University of Illinois, Chicago stressed this urgency in a September 2017 report in the scientific journal *Nature*, noting:

> *We are in a race against time to deploy post-quantum cryptography before quantum computers arrive.*

Bernstein and Lange predict that "Many commonly used cryptosystems will be completely broken once large quantum computers exist."[160]

Are Social Relationships the New Virtual Reality?

Just as information technology and the Internet are impacting our economic and cyber security, they are unquestionably altering our social security with regard to relationships with each other and the world-at-large as well.

Manuel Castellas, professor and chair of Communication Technology at the University of Southern California, observes that these impacts are not entirely either good or bad:

> *People, companies and institutions feel the depth of this [wireless communication] technological change, but the speed and scope of the transformation has triggered all manner of utopian and dystopian perceptions that, when examined closely through methodologically rigorous empirical research, turn out not to be accurate.*[161]

Castellas points out, for instance, that while the media often reports that intense use of the Internet increases the risk of isolation, alienation and withdrawal from society, available evidence shows the opposite. He argues that instead of isolating people and reducing their sociability, "it actually increases sociability, civic engagement, and the intensity of family and friendship relationships, in all cultures."

Posting in Technology Review, Castellas observes that our current "network society" is a product of the digital revolution that has led to major sociocultural changes. One of these is the rise of the "me-centered society" which has both positive and negative aspects. This shift is marked by an increased focus on individual growth, but with an attendant decline in "community" in terms of space, work, family and ascription in general.

Nevertheless, as Professor Castellas explains:

> [I]ndividuation does not mean isolation, or the end of community. Instead, social relationships are being reconstructed on the basis of individual interests, values, and projects. Community is formed through individuals' quests for like-minded people in a process that combines online interaction with offline interaction, cyberspace, and local space.

He continues:

> Today, social networking sites are preferred platforms for all kinds of activities, both business and personal, and sociability has dramatically increased—but it is a different kind of sociability. Most Facebook users visit the site daily, and they connect on multiple dimensions, but only on dimensions they choose. The virtual life is becoming more social than the physical life, but it is less a virtual reality than a real virtuality, facilitating real-life work and urban being.

According to a study conducted by the non-profit Common Sense Media, more than two-thirds of teens say they would rather communicate with their friends online than in person.

The percentage of young people who said their favorite way to talk is face-to-face declined to 32 percent from 49 percent six years earlier (in 2012) according to a survey of more than 1,000 13-to-17-year-olds.

Fifty-seven percent said that social media was a frequent distraction during homework assignments, yet 31 percent said they turn their phones off during all or most homework time. And over 70 percent of those polled believed that tech companies manipulate users to get them to spend more time on their mobile devices.[162]

Manuel Castellas notes, however, that at their root, social networking entrepreneurs are really selling spaces in which people can freely and autonomously construct their lives. Sites that attempt to impede free communication are soon abandoned by many users in favor of friendlier and less restricted spaces. He writes:

> *Messages no longer flow solely from the few to the many, but with interactivity. Now, messages also flow from the many to the many, multi-modally and interactively. By disintermediating government and corporate control of communication, horizontal communication networks have created a new landscape of social and political change.*

These networked online social and political movements, which have been particularly active since 2010, are having broad societal influences. Their impacts are evidenced in the Arab revolutions against dictators, challenges to oppressive state power in Iran and protests against the management of national financial crises throughout the world.[163]

Machine Ethics as a New Religion?

Scientific advancements have led us to thresholds of strange new worlds of contemplation beyond familiar references of human experience. Quantum theory, for example, has uprooted traditional Newtonian views of a Universe where time is linear, gravity "pulls," space has measurable dimensions, or even that a singular "reality" exists outside the influence of our individual thoughts.

As postulated in the Stanford Encyclopedia of Philosophy, information technology either constitutes or is closely correlated with what constitutes our existence and the existence of everything around us including the manner in which the Universe operates. This realization has given rise to the new fields of Information Philosophy and Information Ethics.[164]

Transformational innovations of this revolutionary information era are applying observed, yet poorly understood, principles to create thinking machines with seemingly limitless capacities. Such inventions are already extending—even redefining—the meaning of "artificial intelligence."

If humans can invent machines which are increasingly smarter than we are, where does this lead? Are we in a sense "playing God" in a way that will render human reasoning obsolete? Can we integrate technological "thinking parts" into our biological anatomy to repair and replace failed sensory and motor response systems...just as we presently do with other organ and limb prosthetic devices?

Whatever skeptical views Stephen Hawking and Albert Einstein may have expressed about God and religion, both have delved deeply into mysterious workings of nature at a subatomic level which, to our conventional senses, take on extrasensory, supernatural manifestations.

The "new science" of quantum mechanics goes so far as to suggest a "preposterous" possibility that everything in the physical Universe exists only as illusory inventions of our individual minds. Whereas this concept presents a radical departure from traditional Western thought, it doesn't seem nearly so alien to much older Eastern philosophies.

Generally speaking, whereas Western philosophies tend to emphasize learning new things about what reality is, ancient Hindu and Buddhist literature speaks of removing veils of ignorance that stand between us and what we really are. And where Western religions tend to envision a Universe divided into separate material and spiritual aspects, Eastern teachings make no dichotomous distinctions between material and spiritual manifestations.

Quantum mechanics challenges any notion of material reality altogether, making no distinction between mass (quanta) and their energetic and mysteriously unpredictable relationships with individual observers. In doing so, it has yielded replicable evidence that powers of mind over matter, and realities much stranger than presumed fictions, can no longer be casually dismissed merely as quack clichés.

So let's imagine some possible ethical and moral dilemmas as "beyond material reality" hyper-intelligent systems begin to exert more and more influence over humanity. Documentary filmmaker and author James Barrot, who is excessively alarmist with the premise, stated in the title of his book, *Our Final Invention: Artificial Intelligence and the End of the Human Era.*[165]

Eric Hendry explored Barrot's reasoning behind this ultimate human doom scenario in an interview he posted in the Smithsonian titled *What Happens When Artificial intelligence Turns On Us?*

Barrot begins by reminding us that AI is a dual-use technology that like nuclear fission; it is capable of great good or great harm. He warns that we're just starting to see the harm:

The NSA privacy scandal came about because NSA

> developed very sophisticated data-mining tools. The agency used its power to plumb the metadata of millions of phone calls and the entirety of the internet—critically, all email. Seduced by the power of data-mining AI, an agency entrusted to protect the Constitution instead abused it. They developed tools too powerful for them to use responsibly.

James Barrot then highlights another current ethical battle regarding development and deployment of advanced fully autonomous AI-powered killer drones and battlefield robots that don't have humans in the decision loops. He opines:

> [The policy conflict is] brewing between the Department of Defense and the drone and robot makers who are paid by DoD, and the people who think it's foolhardy and immoral to create intelligent killing machines. Those in favor of autonomous drones and battlefield robots argue that they'll be more moral—that is, less emotional, will target better and be more disciplined than human operators. Those against taking humans out of the loop are looking at drones' miserable history of killing civilians, and involvement in extralegal assassinations. Who shoulders the moral culpability when a robot kills? The robot makers, the robot users, or no one?

Barrot speculates that in the longer term, AI approaching human-level intelligence will neither be easily controlled or possess a benevolent nature towards its creators. Quoting AI theorist Eliezer Yudkowsky of MIRI [the Machine Intelligence Institute]:"The AI does not love you, nor does it hate you, but you are made of atoms it can use for something else."

Reinventing Ourselves

Barrot concludes:

> *If ethics can't be built into a machine, then we'll be creating super-intelligent psychopaths, creatures without moral compasses, and we won't be their masters for long.*[166]

As noted by Barrot, renowned inventor Ray Kurzwell predicted that humans will eventually meld with AI technologies through cognitive enhancements to become "transhumanists." According to this theory, artificial general intelligence will ultimately evolve along with us.

According to Kurzwell, not only will computer implants enhance our brains' speed and overall capabilities, we will also be able to transport human intelligence and consciousness into computers. This strategy would ensure that super-intelligence will be at least partly human, controllable and "safe."

For instance, computer implants will enhance our brains' speed and overall capabilities. Eventually, we'll develop the technology to transport our intelligence and consciousness into computers. Then super-intelligence will be at least partly human, which in theory would ensure super-intelligence was "safe."

Barrot observes that a problem with this theory is that since even we humans aren't reliably safe, it is unwarranted to expect that super-intelligent humans will be either:

> *We have no idea what happens to a human's ethics after their intelligence is boosted. We have a biological basis for aggression that machines lack. Super-intelligence could very well be an aggression multiplier.*

Meanwhile, Barrot laments, 56 nations are developing

battlefield robots, and the drive is to make them, and drones, autonomous. They will be machines that kill, unsupervised by humans.

Impoverished nations will be hurt most by autonomous drones and battlefield robots. Initially, only rich countries will be able to afford autonomous kill bots, so rich countries will wield these weapons against human soldiers of the impoverished nations.

Barrot asks us to imagine that in as little as a decade, a half-dozen companies and nations might field computers that rival or surpass human intelligence:

> *Soon we'll be sharing the planet with machines thousands or millions of times more intelligent than we are. And, all the while, each generation of this technology will be weaponized. Unregulated, it will be a catastrophe.*

George Glider takes strong issue with a "new catastrophe theory" which holds that super-intelligent self-learning machines will make human minds obsolete. He refers to this attitude as "Google Marxism" because Marx's essential theme was that the Industrial Revolution of the 19th century had overcome all the challenges of production. Marx was convinced that the steam turbine, electrification and what William Blake called "dark satanic mills" were the final stage in social evolution—"an eschaton."

Gilder perceives that Google and the Silicon Valley people also imagine that their artificial intelligence, their machine learning, their cloud computing, is an eschaton—another "end of history moment."

Gilder contends that Google ideology believes in a "winner-take-all" Universe where the age of capitalism is at an end. They see the existing generation of capitalists as the final capitalists. "That's their vision." He argues that if you believe that "machines can re-create new machines in a steady cascade of greater capabilities that are beyond human comprehension and control, you really believe that's the end of the human race."[167]

This narrative of human obsolescence gives rise to Google's proposal that the only way forward is to end the market economy by providing citizens some sort of "guaranteed annual income." Glider warns:

> *If everyone gets supported without any kind of growing up and facing the challenges of life, then our capitalist culture would collapse.*

Glider argues that too much negativity is being expressed in the American media and on college campuses regarding "the dangers and perils of [information] technology rather than its promise." He emphasizes that the very nature of information is surprise, just as human creativity always comes as a surprise. "If it wasn't surprising, we wouldn't need it."

Human minds can generate counterfactuals, imaginative flights and dreams. By contrast, Glider says, "a surprise in a machine is a breakdown. You don't want your machines to have surprising outcomes!"

Bioengineering our Human Society

Michael Bess, Vanderbilt University professor and author of *Our Grandchildren Redesigned: Life in a Bioengineered Society,* foresees a human future that is both terrifying and promising. In an interview with Vox.com contributor Sean Illing, Bess raises special concerns regarding AI's social influences on society and its ultimate potential to enable biological reengineering of our lives altogether.[168]

Bess acknowledges that people have panicked about

impacts of new information technologies since the invention of the printing press.

Even long before that, Socrates argued that reading a manuscript was nowhere near as insightful as talking with its author:

> *[Written words] seem to talk to you as though they were intelligent, but if you ask them anything about what they say, from a desire to be instructed, they go on telling you the same thing forever.*[3]

Now, the advent of smartphones, computers and the Internet seem to be comparable in their impact to other big revolutions in communications and transportation that we've experienced over the past thousand years.

Bioengineering, however, is different. The impact of social media will pale in comparison to potential revolutions in AI or gene editing technologies. Bess projects that we're now on the verge of developing DNA-altering technologies that are so qualitatively different and more powerful that they will force us to reassess what it means to be human:

> *Bioelectric implants, genetic modification packages - the ability to tamper with our very biology—this stuff goes far beyond previous advances, and I'm not sure we've even begun to understand the implications.*

What's more, such capabilities are advancing at an unprecedented rate. Bass observes:

> *We went from having no World Wide Web to a full-blown World Wide Web in 20 or 25 years—that's*

---

[3] *Phaedrus*, section 275d

astonishing when you consider how much the Internet has changed human life. In the case of, say telephones that took many decades to fully spread and become as ubiquitous as it is today.

So what we've seen with the Internet is blisteringly fast compared to the past. For most of human history, the world didn't change all that much in a single lifetime. That's obviously not the case anymore, and technology is the reason why.[169]

Bess worries that mankind doesn't have enough time to adapt to these changes…adequate time to alter our habits and to reappraise our cultural sense of who we are:

When these things happened slower in previous eras, we had more time to assess the impacts and adjust. That is simply not true anymore. We should be far more worried about this than we are.

Sean Illing observed that our technology is developing much faster than our culture and institutions, and that this growing gap will eventually destabilize society. Here, Bess was less pessimistic:

I think overall, as a society, we're insufficiently equipped, but that doesn't mean there aren't plenty of voices out there speaking sanity. What's interesting is that you can use these new technologies to get in touch with those voices and connect with other people who are questioning these technologies. The ability to connect in that way offers a lot of promise if it's used wisely.

Professor Bess added that while many young people appear to be walking around college campuses mindlessly staring at their phones, it's clear that even they understand what's happening and why it's problematic:

> *The more you live through screens, the more you're living in a narrow bandwidth, an abstract world that's increasingly artificial. And that virtual world is safe and controllable, but it's not rich and unpredictable in the way the real world is. I'm worried what will happen if we lose our connection to reality altogether.*

What's most striking about us humans, Bess observes, is that we are unpredictable in very basic ways. We're more complex than we can fathom, and there's something about us that is the opposite of artificial. It's the opposite of something made:[170]

> *All this genetic modification technology has the potential to take us into very worrisome territory where all the things we hold dear in our current world, all the values that give our lives meaning, are at risk. Either our survival is at risk or we become semi-machines who are like the marionettes of our own moment-to-moment experience. What becomes of autonomy? What becomes of free will? All these questions are on the table.*

Bess urges each of us to ask ourselves:

> *What does it mean for a human being to flourish? These technologies are forcing us to be more deliberate about asking that question. We need to sit down with ourselves and say, "As I look at my daily life, as I look*

> at the past year, as I look at the past five years, what are the aspects of my life that have been the most rewarding and enriching? What things have made me flourish?

Bess concludes:

> If we ask these questions in a thoughtful, explicit way, then we can say more definitely what those technologies are adding to the human experience and, more importantly, what they're subtracting from the human experience.[171]

Ratcheting up the bioengineering potentials even farther, what if artificial intelligence begets artificial life? Such an idea is no longer an implausible script of science fiction fantasy
 As explained in the *Stanford Encyclopedia of Philosophy*, artificial life (Alife) is an outgrowth of AI technology to simulate or synthesize life functions:

> The problem of defining life has been an interest in philosophy since its founding. If scientists were to succeed in discovering the necessary and sufficient conditions for life and then successfully synthesize it in a machine or through synthetic biology, then we would be treading on territory that has significant moral impact.[172]

One form of ALlife which was inspired by the work of mathematician John von Neumann aims to achieve computational models which produce self-replicating cellular automata called "Loops." So far, these ALlife applications are content to create programs that simulate life functions rather than demonstrate "intelligence." A primary moral concern here is that these programs are designed to self-reproduce and in a way that resembles computer viruses. Successful Alife programs can potentially become computer malware vectors.

A second form of ALlife is based upon manipulating actual biological and biochemical processes in such a way that it produces novel life forms not seen in nature. It is much more morally charged.

In May 2010, scientists at the J. Craig Venter institute were able to synthesize an artificial bacterium called JCVI-syn1.0. Referred to as "Wet ALlife," this development tends to blur boundaries between bioethics and information ethics, potentially leading to dangerous bacteria or other disease agents, just as software viruses infect computers.[173]

Some even argue that information is a legitimate environment of its own which possess intrinsic value that is in some ways similar to the natural environment, while in other ways foreign. Either way, they propose that information, as is on its own a thing, is worthy of ethical concern.

Therefore, if AI-directed robots are information machines, is it ethical to unplug and virtually kill them? Would HAL in *2001 Space Odyssey* really have cared?

Probably not.[174]

Are We Ready for Posthumanism?

It's tempting to imagine that the self-learning AI-powered intellects we create that surpass our human mental capacities would still be much like us in important respects. They would just be a lot cleverer. The same hope might prevail as we humans, with the help of smart implants and bioengineered artificial DNA, evolve (or perhaps devolve) into a new posthuman variant of our current precursor model.

If so, what about those nostalgic things we may trade in for the upgraded version? Things we tend to value such as capacities for love and compassion, experiencing inspiration and joy, even mortal survival? Such characteristics are all shared with histories that have given rise to evolutions of other "higher animals"—but not with computer program prodigies such as Hal.

The good news here might be that while intelligent machines and updated versions of ourselves might not share our current values, they might also lack tendencies towards hostility, another frequent expression of animal emotion.

The bad news is that those emerging creations might simply be as indifferent to us as we are to bugs on our car windshields.[175]

Mary Midgley, a researcher at the Newcastle University Royal Institute of Philosophy in the United Kingdom and author of *Philosophical Plumbing* argues that the belief that science and technology will bring us immortality and bodily transcendence is based on pseudoscientific beliefs premised upon a deep fear of death (apparently HAL was an exception).[176]

John Sullins, a professor of philosophy at Sonoma State University in California, argues that there is a quasi-religious aspect to the acceptance of transhumanism, and that the transhumanist hypothesis influences the values embedded in computer technologies are dismissive or hostile to the human body.

While many ethical systems place a primary moral value on preserving and protecting the natural, Sullins criticizes transhumanists for rejecting value distinctions in defining what is natural and what is not. Instead, they tend to dismiss arguments to preserve some perceived natural state of the human body as antithetical to progress.[177]

Philosopher Nick Bostrom at the Future of Humanity Institute at Oxford University argues that putting aside the feasibility argument, we must conclude that there are forms of posthumanism that would be beneficial, such as enabling longer, more worthwhile lives. He favors such prospects if at all possible.[178]

Writers Huw Price and Jean Tallinn ask in their Conversation article blog if it is necessarily a bad thing if computers become as smart, or smarter, than humans. They point out that the narrow list of AI application successes so far is mostly useful.

They observe:

> A little damage to Grandmasters' egos, perhaps, and a few glitches on financial markets, but it's hard to see any sign of impending catastrophe...[179]

Price and Tallinn believe that greatest AI concerns stem from the possibility that computers might take over domains that are critical to controlling the speed and direction of technological progress itself. They ask: "What happens if computers reach and exceed human capacities to write computer programs?"

The first person to consider this possibility was Cambridge-trained mathematician IJ Good. In 1965 he observed that having intelligent machines develop even more intelligent machines would leave the human levels of intelligence far behind. He called the creation of such machine "our last invention."

In this scenario, the moment computers become better programmers than we do will mark a point in history where the speed of technological progress shifts from the speed of human thought and communication to the speed of silicon.

This is a version of Vernor Vinge's technological singularity—where beyond this point, the curve is driven by new dynamics and the future becomes radically unpredictable. The former San Diego State University mathematical science professor wrote that whether or how humans will be part of future superintelligence—or what it will ultimately mean to people, cannot presently be known. He compared attempting to explain the technology developed in the posthuman future of artificial intelligence as being like trying to explain *Plato's Republic* to a mouse.[180]

Price and Tallinn argue that even if we don't accept the premise of a looming technological singularity—that theoretical point at which AI will outstrip all human intelligence—AI will continue to play increasingly central roles in our everyday lives. Whereas various authorities in the software arena continue to debate the possibility of a "strong AI" (artificial intelligence that matches or exceeds human intelligence), a large caravan of "narrow AI" (AI that's limited to particular tasks) races steadily and speedily forward.

One by one, computers are taking over domains that were previously considered off-limits to anything but human intellect and intuition.

The late Stephen Hawking has warned that artificially intelligent machines could even kill us when they become too clever. Responding to a question during his first Ask Me Anything session on Reddit, he said:

> *The real risk with AI isn't malice but competence. A super intelligent AI will be extremely good at accomplishing its goals, and if those goals aren't aligned with ours, we're in trouble.*
>
> *You're probably not an evil ant-hater who steps on ants out of malice, but you're in charge of a hydroelectric energy project and there's an anthill in the region to be flooded, too bad for the ants. Let's not place humanity in the position of those ants.*

Hawking proposed that there is no limit to what human intelligence can create: [We] evolved to be smarter than our ape-like ancestors, and Einstein was smarter than his parents.

He predicts that our human AI self-learning inventions will lead to "machines whose intelligence exceeds ours by more than ours exceeds that of snails."

How soon might this happen? Hawking said he had no idea, and warned not to trust "anyone who claims to know for sure that it will happen in your lifetime or that it won't happen in your lifetime. But when it happens, "it's likely to be either the best or worst thing ever to happen to humanity, so there's huge value in getting it right."

As such, Hawking urged that we must "shift the goal of AI from creating pure undirected artificial intelligence to creating beneficial intelligence." [181]

James Barrot agrees that while an obvious goal would be to imbue super smart AI technologies with a moral sense that makes them value human life and property, programming ethics into a machine turns out to be extremely hard because norms change over time and are contextual. He asks:

> *If we humans can't agree on when life begins, how can we tell a machine to protect life? Do we really want to be free? We can debate it all day and not reach a consensus, so how can we possibly program it?*[182]

On the other hand, if we fail, we may only succeed in creating super-intelligent psychopaths, creatures without moral compasses, and we won't be their masters for long.

At the same time, Price and Tallinn point out that people sometimes complain that the same concern applies to corporations as psychopaths when they aren't sufficiently reined in by human control. The pessimistic prospect here is that AI might be similar, except much, much, cleverer, and also much faster.[183]

As Victor Vinge concluded, all predictions about our

ultimate human AI future, enthusiastic or not, are destined to be inconclusive. Yet just as survival in our here-and-now lives requires taking uncertainties very seriously when a lot is at stake, that future demands no less.

Price and Tallinn offer some good advice in this regard. They write:

> *A good first step, we think, would be to stop treating intelligent machines as the stuff of science fiction, and start thinking of them as a part of the reality that we or our descendants may actually confront, sooner or later.*

Price and Tallinn optimistically conclude that the future isn't yet fixed, but that such optimism will only be warranted if we take the trouble to make the future an optimistic one by investigating the issues and thinking hard about the safest strategies:

> *We owe it to our grandchildren—not to mention our ancestors, who worked so hard for so long to get us this far—to make that effort.*[184]

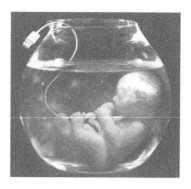

Human Evolution And Revolution: Trends, Triumphs and Trepidations

IN MARY SHELLEY'S famous 1818 novel, Dr. Frankenstein created intelligent life out of inanimate matter—and unfortunately too late, regretted meddling with nature. Will this same lesson prove to be the case with AI? Will our human story end in tragedy of Frankenstein proportions, or will we be able to live in harmony?

After all, much like many view AI today, even though the monster possessed moral and emotional sensibility, society unfairly and violently rejected its appearance and strength out of fear.

Despite good intentions and deeds, the poor creature just couldn't seem to win public support. As the monster described himself, "my life has been hitherto harmless and in some degree beneficial." He even used "extreme labour" to rescue a young girl from drowning, but no matter what he did, those actions were always misinterpreted. The public assumes that he was trying to murder the girl, and William Frankenstein even assumes that his monstrous creation plans to kill him.

Perhaps immodestly, the monster had a very good opinion of his superiority over his mortal detractors. He said:

> *I was not even of the same nature as man. I was more agile than they and could subsist upon coarser diet; I bore the extremes of heat and cold with less injury to my frame; my stature far exceeded theirs.*

Mary Shelley, the monster's real-life creator, understood our natural tendency to fear what we do not understand. She wrote: "Nothing is so painful to the human mind as a great and sudden change." As Victor Frankenstein lamented:

> *I started from my sleep with horror; a cold dew covered my forehead, my teeth chattered, and every limb became convulsed: when, by the dim and yellow light of the moon, as it forced its way through the window shutters, I beheld the wretch—the miserable monster whom I had created.*

Nevertheless, that fearsome creature conjured by Mary's imagination warned that unfairly pre-judged resistance to change would portend dire consequences. The monster cried out:

> *Shall each man find a wife for his bosom, and each beast have his mate, and I be alone? I had feelings of affection, and they were requited by detestation and scorn.*
>
> *Man! You may hate, but beware! Your hours will pass in dread and misery, and soon the bolt will fall which must ravish from you your happiness forever. Are you to be happy while I grovel in the intensity of my wretchedness?*
>
> *You can blast my other passions, but revenge remains—revenge, henceforth dearer than light or*

> *food! I may die, but first you, my tyrant and tormentor, shall curse the sun that gazes on your misery. Beware, for I am fearless and therefore powerful.*

Having been warned that acting against the monster's wishes would cause him to lose everything, including his good reputation, Victor recognizes that the danger to the world is greater than consequences to himself. Accordingly, he chooses to sacrifice himself to atone for his hasty rush into scientific inquiry.

Frankenstein-like theories regarding fears and fortunes of AI-driven monsters are subjects of contentious debate in today's scientific, technological, philosophical and public policy communities. Here, as in the past, our attention drifts to extremely contrasting and divided visions which are most dramatic rather than most likely.

One of the best-known members of the dystopian camp, Elon Musk, has called Super-intelligent AI systems "the biggest risk we face as a civilization," comparing their creation to "summoning the demon." Some sharing his view warn that when humans create self-improving AI programs whose intellect dwarfs our own, we will lose the ability to understand or control them.

Utopians, on the other hand, are more inclined to expect that once AI far surpasses human intelligence, it will provide us with near-magical tools for alleviating suffering and realizing human potential. Some holding this vision foresee that super-intelligent AI systems will enable us to comprehend presently unknowable vast mysteries of the Universe, and to solve humanity's most vexing questions such as eradication of diseases, natural resource depletion and world hunger.

Both of these scenarios would require that our AI developments lead to "artificial general intelligence" which can handle the incredible diversity of tasks accomplished by the human brain. Whether or not this will ever happen, much less how those tasks will be transformed and when, remain to be pure conjecture.

A Mixed Bag of Prophases

A certain fact remains that AI, automation and the Internet are already impacting our lives in large and small ways that are legitimately argued as both good and bad. Moreover, the progeny of this triumvirate of Frankenstein monsters will continue to multiply to exert more and more influence over ever-broader aspects of our lives at an accelerating rate.

Lacking the apocalyptic drama of Hollywood blockbusters, some inherent perils might be likened by to a "boiling frog" analogy. Like a hapless frog placed comfortably in an open container of tepid water which is then brought slowly to a boil, it will not perceive danger until it is too late to jump out and is cooked to death.

Although gradual, there is broad recognition that smart technology will increasingly disrupt traditional structures of our social lives and economic livelihoods. Key among these impacts is a looming AI-driven work displacement crisis which will dramatically widen the wealth gap and pose a broad-spread challenge to maintenance of personal dignity.

Society has trained most of us to tie our personal worth to the pursuit of work and success. It will be painful for those who watch algorithms and robots replace them at tasks they have spent years mastering and proudly attending. Many will witness those tasks and entire industries disappear altogether, as ill-fated buggy whip manufacturers experienced following the invention of the internal combustion engine.

Joel Mokyr recognizes that while AI and automation are boosting economic growth by creating new types of jobs and improving efficiency in many businesses, they will be accompanied by negative effects on others. In the near-term, the less educated workers are likely to represent a disproportionate percent of job-loss casualties.[185]

Dr. Kai-Fu Lee, author of *AI Superpowers: China, Silicon Valley and the New World Order,* argues that unprecedented disruptions applying existing AI technology to new problems will hit many white-collar professionals just as hard as it hits blue-collar factory workers. Still, he says:

> *Despite these immense challenges, I remain hopeful. If handled with care and foresight, this AI crisis could present an opportunity for us to redirect our energy as a society to more human pursuits: to taking care of each other and our communities. To have any chance of forging that future, we must first understand the economic gauntlet that we are about to pass through.*[186,187]

Lee points out that techno-optimists and historians would argue that productivity gains from new technology almost always produce benefits throughout the economy, creating more jobs and prosperity than before, but with a mixed bag of impacts. He notes:

> *The steam engine and electrification created more jobs than they destroyed, in part by breaking down the work of one craftsman into simpler tasks done by dozens of factory workers. But information technology (and the associated automation of factories) is often cited by economists as a prime culprit in the loss of U.S. factory jobs and widening income inequality.*[188]

Lee concludes his article optimistically:

> Artificial Intelligence will radically disrupt the world of work, but the right policy choices can make it a force for a more compassionate social contract.

Will Tech Overlords Lord Over All?

Those entities which control AI and information technologies will determine employment winners and losers. Consequently, critical uncertainties regarding ultimate threats posed by this new AI Frankenstein monster may revolve less around our mastery of its invention, or its mastery over us, and far more about who will ultimately master control over both.

Companies with more data and better algorithms will gain ever more users and data. This monopolistic self-reinforcing winner-take-all cycle will lead to God-like controls over all segments of society unknown in human history. Their instruments of power include dominion over information access and censorship, individual and business privacy, physical and economic security, transportation and energy infrastructures and financial levers of political influence which are growing at an astounding rate.

Recent events portend frightening global social consequences of amassing enormous quasi-government-level concentrations of wealth and power in a handful of monopolistic enterprises.

After concerns about Google's corporate practices lead to a revolt among the company's programmers, its CEO, Sundar Pichai, outlined new corporate guidelines for ethical principles. Issuing a blog post, Pichai wrote that Google would not produce:

- Technologies that cause or are likely to cause overall

harm.

- Weapons or other technologies whose principal purpose is to cause or directly facilitate injury to people.

- Technology that gathers or uses information for surveillance violating internationally accepted norms.

- Technologies whose purpose contravenes widely accepted principles of international law and human rights.

Pichai also laid out an additional seven principles to guide the design of future AI systems:

- AI should be socially beneficial.

- It should avoid creating or reinforcing bias.

- It should be built and tested for safety.

- It should be accountable. Incorporate privacy design principles.

- It should uphold high standards of scientific excellence.

- It should be made available for use.[189]

Google's ongoing behaviors directly contradict much of that lofty rhetoric.

In a speech on October 1, 2018 at the Hudson Institute, a conservative think tank focused on security and economic issues, Vice President Pence called on U.S. companies to reconsider business practices in China that involve turning over intellectual property or "abetting Beijing's oppression." He said, "For example, Google should immediately end development of the Dragonfly app that will strengthen Communist Party censorship and compromise the privacy of Chinese customers."[190]

Mr. Pence's speech was the first public White House condemnation of Dragonfly, a mobile version of Google's search engine which is being designed and tested to adhere to China's strict citizen censorship program.

A spokeswoman for Google, a unit of Alphabet Inc. who declined to comment about the criticism, simply referred to a previous statement that described the company's work as exploratory and "not close to launching a search product in China." A logical follow-up question would be, "exploratory to what purpose?"

That purpose would appear to be quite obvious. During the same talk, Pence accused China of seeking to "foster a culture of censorship" in academia. In doing so the vice president cited a speech from Yang Shuping, a University of Maryland student from China who became a target of criticism there after praising the "fresh air of free speech" in America. Ms. Yang became the victim of a firestorm of criticism on China's tightly controlled social media, and her family was harassed as well.

Google's social media censorship collaboration with China becomes ever more troubling as that country now aggressively extends its weaponization of information technology capabilities into cyberspying beyond their own borders. New Chinese cybersecurity rules give their authorities sweeping powers to inspect companies' information technology and access proprietary information—steps that warrant deep concern of all foreign businesses dealing with China operations.[191]

Chinese police officials are now authorized to remotely access corporate networks for potential security loopholes, to copy information and to inspect any records that "may endanger national security, public safety and social order." Those categories provide carte blanche opportunities to be interpreted any way they are deemed useful.

The cybersecurity law also mandates internal security checks on technology products that are supplied to the Chinese government and to critical industries such as banking and telecommunications. Foreign companies operating in China will be held responsible for allowing prohibited information to circulate online, and Internet service providers must provide "technical support" to authorities during national security or criminal investigations.

According to William Nee, an analyst at Amnesty International:

> *[The new ruling] authorization strengthens the state's authority to inspect and requires that Internet-service providers and companies using the Internet are fully complying with the government's cybersecurity prerogatives.*

In addition, the law mandates that foreign companies working in China store their data in that country. This requirement enables Beijing to force the disclosure of source codes and other corporate secrets, forcing companies to prove their equipment is secure. The Chinese government can then potentially leak the information to domestic competitors.

As the *Wall Street Journal* reports:

> *The rules will do nothing to assuage foreign companies' worries about security of their property. They grant authorities access to any information related to cybersecurity—a category so broadly defined as to include just about everything.*

William Zarit, chairman of the American Chamber of Commerce in China, said:

> *It justifies for authorities the right to basically copy or access anything. It doesn't seem like companies have a*

choice.

In dutiful compliance, Microsoft Corp. has opened what it calls a "transparency center" in Beijing where officials can test its products for security. Apple Inc. has started building a data center in the province of Guizhou to comply with rules requiring cloud data from Chinese customers to be stored in China.

In anticipation of the law, Amazon transferred operational control of its Beijing data center to its local partner, Beijing Sinnet. As the *Wall Street Journal* reported:

> The following November, Amazon sold the entire infrastructure to Beijing Sinnet for about $300 million. The person familiar with Amazon's probe casts the sale as a choice to 'hack off the diseased limb.' [192]

Cyber Attacks on Privacy and Security

As reported by *Bloomberg Newsweek*, in 2015, Amazon began quietly evaluating a startup called Elemental Technologies, a potential acquisition to help with a major expansion of its streaming video service, known today as Amazon Prime Video. Elemental made software for compressing and formatting massive video files for different devices.[193]

Amazon Web Services (AWS) was interested at the time in using Elemental Tech for highly secure cloud servers being assembled for them by Super Micro Computer Inc. (Supermicro), a San Jose, California-based company...one of the world's biggest suppliers of server motherboards, the fiberglass-mounted clusters of chips and capacitors that act as the neurons of data centers large and small.

According to *Newsweek*, AWS was surprised to discover tiny

microchips about the size of a grain of rice nested in some of the server motherboards that didn't belong there. Amazon reported this to find authorities—bugs that had the potential to end up in the International Space Station, in DOD data centers, controlling CIA drone operations and onboard networks of Navy warships.

The insidious little chips provided hackers with stealth doorways into any network that included other altered machines—all of which had a common connection. The malware all came from China.

Since the implants are tiny, the amount of code they contain is small. Nevertheless, they are capable of doing two important things: They can tell the device to communicate with one of several anonymous computers elsewhere on the Internet that are loaded with a more complex code, and they can also prepare the device's operating system to accept this new code.

The Illicit chips could do all this because they were connected to the baseboard management controller, a kind of superchip that administrators use to routinely log in to problematic servers. This gives them access to the most sensitive code, even on machines that have crashed or are turned off.

Even more, the spy chips can also alter part of that code so that the server won't check for a password, opening it up to all users. This further enables them to steal encryption keys for secure communications, to block security updates that would neutralize the attack, and to open up new pathways to the Internet.

Machines can be compromised to invite security breaches in two ways. One (interdiction) involves manipulating devices while they're in transit from manufacturer to customer. The other involves seeding changes from the beginning. China has big advantage here, since they make an estimated 75 percent of the world's mobile phones and 90 percent of PCs.

That's exactly what AWS inspectors reportedly found had

happened. Two company officials reported that the illicit chips were traced to operatives from a unit of the People's Liberation Army and four Chinese computer part subcontracting companies.

The stealth chips were devised to be as inconspicuous as possible. In gray or off-white colors they looked more like signal conditioning couplers, a common motherboard component. Slight variances in size suggested that they came from different factories.

The corrupted Supermicro computers reportedly wound up in the possession of nearly 30 companies, including a major bank and government contractors. Apple, Inc., an important Supermicro customer, started removing all of its 7,000 Supermicro servers from its numerous data centers following the discovery.

Amazon denied the claims. They wrote:

> *On this we can be very clear: Apple has never found malicious chips, 'hardware manipulations,' or vulnerabilities purposely planted in any server.*

A spokesman for Supermicro, Perry Hayes, wrote, "We remain unaware of any investigation." The FBI and Office of Director of National intelligence, representing the CIA and NSA, declined to comment.

Newsweek asserts that the companies' denials were countered by six current and former senior national security officials, who—in conversations that began during the Obama administration—detailed the discovery of chips and the government investigation.

However, no consumer data is known to have been stolen.

According to *Newsweek*, the stealth chips discovered in Supermicro devices were crude compared with far more sophisticated designs an Amazon security team found in altered motherboards being assembled in AWS's Beijing facilities. In one case, the malicious chips were thin enough that they'd been embedded between the layers of fiberglass onto which the other components were attached. One person who saw pictures of the chips said that this more advanced generation of chips was smaller than a sharpened pencil tip.

(Again, *Newsweek* said that Amazon denies that AWS knew of servers found in China containing malicious chips.)[194]

Supermicro was delisted from the Nasdaq stock market trading in 2018.

Reuters World News reported in December 2016 that the FBI was then investigating how hackers had infiltrated computers at the Federal Deposit Insurance Corporation (FDIC), one of three federal agencies that regulate commercial U.S. banks, over several years beginning in 2010. Internal communications referred to the attacks as having been carried out by Chinese military-sponsored hackers.

The FDIC reported to Congress at least seven "major" cybersecurity incidents during 2015 or 2016.

FDIC spokeswoman Barbara Hagenbaugh declined to comment on the previously unreported FBI investigation or the hack's suspected link to the Chinese military, but said that the regulator took "immediate steps" to root out the hackers when the security breach was discovered.[195]

In late September 2015, President Obama held a White House press conference with Chinese President Xi Jinping announcing a joint understanding that America would no longer support the theft of intellectual property to benefit Chinese companies. *Newsweek* reported that China was willing to offer this concession because it was already developing far more advanced and surreptitious forms of hacking founded upon its near monopoly of the technology supply chain.

Just weeks following the announcement, the Obama administration quietly raised alarm with several dozen tech executives and investors at a small, invite-only meeting organized by the Pentagon. According to an unnamed participant, Defense department officials briefed the technologists about a recent cyberattack and asked them to think about creating commercial products that could detect hardware implants. *Newsweek* reported:

> *Attendees weren't told the name of the hardware maker involved, but it was clear to at least some in the room that it was Supermicro, the person says.*

Newsweek noted that few companies have the resources of Apple and Amazon, and it took some luck even for them to spot the problem:

> *This stuff is at the cutting edge of the cutting edge, and there is no easy technological solution. You have to invest in things that the world wants. You cannot invest in things that the world is not ready to accept yet.*[196]

Smart Technology Lapdogs and Luddites

Which technological solutions "the world" is willing to accept will continue to depend a lot upon individual world perspectives viewed from diverse and divergent geographical, social, economic, ideological and lifestyle-oriented vantage points.

Some will more readily accept the China model, premised upon wholesale trades of personal privacy in exchange for "higher" social benefits afforded by public security, improved energy and infrastructure efficiencies and special privilege incentives awarded to individuals based upon government-dictated and monitored good behavior merits. This model might portend rather grim "smart cities" trending.

Others will be less likely to embrace wired-together metropolitan convenience vs. losses of privacy. Individual autonomy tradeoffs may increasingly opt for more suburban and rural lifestyles afforded by Internet-connected remote entrepreneurship and telecommuting opportunities. This trend, in combination with the former, will likely deepen ideological and political demographics nationwide, and regional and local divides.

A 2017 *Washington Post Family Foundation* opinion poll revealed substantial lifestyle priority and political differences between Americans living rural settings versus metropolitan centers. A key finding was that this divide is more cultural than economic in nature, and these convictions were held most strongly in rural communities.[197]

The survey of nearly 1,700 Americans—including more than 1,000 adults living in rural areas and small towns—found a deep-seated kinship in rural America, coupled with a stark sense of estrangement from people who live in urban areas. Nearly 7 in 10 rural residents said that their values differed from those of people who live in big cities, including 4 in 10 who said that their values are "very different." By comparison, about half of the urban residents say their values differ from rural people, with less than 20 percent of urbanites believing that rural values are "very different."

Rural misgivings regarding metropolitan attitudes included predominate concerns about America's rapidly changing demographics, a sense that traditional Christian values are under siege and a perception that the federal government primarily caters most to the needs of people in big cities.

The survey showed that key differences between rural and urban America ultimately center on fairness, questions regarding: "Who wins and loses in the new American economy, who deserves the most help in society and whether the federal government shows preferential treatment to certain types of people."

When asked which is more common—that government help tends to go to irresponsible people who do not deserve it or that it doesn't reach people in need—rural Americans are more likely than others to say they think people are abusing the system.

Rural Americans were also broadly skeptical that the federal government is fair or effective at improving people's economic situations. More than 60 percent said that federal efforts to improve living standards either make things worse or have little impact. Those views appear to feed the rural-urban divide: A 56 percent majority of rural residents says the federal government does more to help people living in and around large cities, while 37 percent feel they treat both urban and rural areas equally.

Rural and metropolitan survey respondents both expressed widespread concerns about the lack of jobs in communities. Two-thirds of rural residents rated local job opportunities as fair or poor, compared with about half of urban residents. Nearly 6 of 10 rural residents said that they would encourage young people in their community to leave for more opportunity elsewhere.

According to Census Bureau data, the poverty rates are similar, 16 percent in cities and 17 percent in rural areas. However, as discussed later, these statistics, based entirely on "money income," fail to account for a variety of value-laden government benefit transfers to low-income recipients and are very misleading. Taken altogether, it is difficult to quantify the extent to which these de facto federal government poverty subsidies work to the special advantage of either demographic set addressed in the poll.[198]

The *Washington Post Family Foundation* opinion poll results highlighted deepening conservative versus liberal political divisions between rural and urban Americans. While urban counties favored Hillary Clinton by 32 percentage points in the 2016 election, rural and small-town voters backed Trump by a 26-point margin, significantly wider than GOP nominee Mitt Romney's 16 points four years earlier.

Most rural residents polled said that they believed that key elements of President Trump's economic agenda would help their local economy. Large majorities attributed this optimism to infrastructure investments, better foreign trade deals, a crackdown on undocumented immigrant workers, lower business taxes and government deregulation.

The survey responses, along with follow-up interviews and focus groups the pollsters conducted in rural Ohio, brought into view a portrait of a split that is tied even more to social identity than to economic experience.

A vast majority of rural Americans judged their communities favorably as places where people look out for each other, which was cited as a point of pride and distinction they say they can't find in urban centers. One follow-up survey responder, Ryan Lawson, who grew up in northern Wisconsin, was quoted as saying:

> Being from a rural area, everyone looks out for each other. People in my experience, in cities are not as compassionate toward their neighbor as people in rural parts.

Lifestyle Matters

So what influences and impacts will rapid advancements and expanding applications of smart technology have upon these economic, social and political divides? As predicated upon rationale that follows, I foresee that they will become increasingly contentious.

First, the Internet now affords much greater opportunities

for people to freely choose where they live. More and more who work for large metropolitan-based corporations will telecommute rather than physically travel on highways or public transportation. This will benefit companies by reducing their need to provide costly metropolitan staff facilities. They'll be able to expand their talent pool on a needs basis without regard to geographic considerations.

Second, many of those Internet-connected corporate service providers, along with other independent entrepreneurial businesses both large and small, can reside wherever their lifestyle choices lead them. Those decisions will take a variety of considerations into account, very much including location-based cost of living values, family-based community assessments, access to preferred natural and cultural settings, ties to familiar and cherished social structures and of course, availability of essential revenue opportunities.

Third, and as previously discussed, AI and automation will disproportionately impact some work and employment markets both to the benefit and detriment of others. While we can expect cities, suburbs and rural areas to win some and lose some competitive advantages, each of us are likely to bring our personal life experiences and biases into any speculations regarding true winners and losers. Here, I plead no exception.

Fourth, let's perhaps assume, as a general rule, that many work and employment opportunities that have traditionally involved metropolitan-based specialized expertise, law and medicine, for example, are displaced by AI servants, and that simultaneously, large metropolitan-centered hospitals, for example, are supplanted by smaller, more broadly distributed AI-supported community clinics—at suburban and rural locations with greater abundances of more affordable personal healthcare workers.

Finally, in the large scheme of things, let's imagine very different demographically-influenced responses to various government policy proposals premised upon remediating "income-inequality" impacts of labor-intensive job losses to automation. Some highly industrial cities, along with suburban and rural communities, particularly those in less prosperous locales, will have greater stakes than others.

Broadly viewed, it's popularly accepted that cities, suburbs and the rural countryside all offer different living experiences which various people prefer. For example, a city dweller is likely to place a higher value on easier access to more business, entertainment and museum/theater culture than someone who lives in a rural area. The rural countryside lifestyle is far cheaper than living in a city, and some may find themselves driving hours to get to a movie theater or a grocery store.

Some differences, including those with political roots and overtones, while distinct, become exaggerated, stereotypic and even humorous. As Americans, just as citizens of other nations, we share a common history, common cultural and family values, common economic needs and self-fulfillment aspirations and a common future. We celebrate our independence to speak and live freely, and we celebrate our enrichment and empowerment afforded by non-prejudicial diversity.

To a substantial degree, we Americans exercise our wonderful freedoms through residency lifestyle preferences relative to income requirements. Consider, for example, that the average 2017 monthly rental cost for a two-bedroom Manhattan apartment was $3,422, although admittedly this is an extreme case.[199]

Smaller cities typically offer much cheaper rental prices, property costs and real estate taxes in widely differing ranges. Even in New York City, rent in the East Village is far pricier than in the South Bronx.

Overall, the average monthly rent in U.S. cities and suburbs was $1,640, with suburbs averaging $1,695—just $50 more expensive. Although luxury homes in the country can also be very expensive, typical rural housing prices are much more affordable than those in urban or suburban communities.

Whereas large cities consist of many thousands or millions of people living in close proximity and set a costly premium on personal space, suburban settings offer a balance; three and-four-bedroom houses are common, as are houses with multiple stories, basements and garages. Nevertheless, many young and older people everywhere will continue to be content with life in small studio apartments with a roommate or as single-person lifestyles.

Families more typically prefer a home with a reasonable amount of outdoor space for kids and pets. Rural living expands affordable yard space, affords more privacy and extends more immediate access to the natural landscape, family and individual outdoor recreation activities.

Transportation requirements, options and economies will continue to be different in urban, suburban and rural communities over the foreseeable future. Some circumstances will continue to depend heavily upon riding trains, subways and buses.

Suburban areas, depending upon size and proximity to metropolitan population centers, will continue to represent various mixes of private vehicle and public transportation connections. Small towns, and rural areas most particularly, will continue to rely predominantly upon private vehicles.

So in balance, while rural and suburban dwellers may spend less on housing, many—depending again on specific location—may wind up spending more on transportation. On the other hand, country living frees residents from having to worry about buying public transportation passes, scheduling activities around transportation and the hassles of transferring between bus stops and subway stations.

City, suburban and rural settings offer very different sorts of business, activity and lifestyle attractions. Broadly generalized, cities are stereotypically characterized as fast-paced "something is always happening" and socially "up-scale" environments that appeal to aggressive "go-getters." At the same time, they are also places inhabited by a much larger majority of people whose fast pace involves running to scrape by, along with those who have given up the race altogether.

Suburbs are broadly regarded as good and safe places to raise children surrounded by other families with shared community priorities. Prized attractions and amenities include close access to active businesses, convenience shopping and community services and high standard schools.

Rural lifestyles are in many ways opposite of urban. There's still a lot to do, but at a slower, more relaxed pace with a less distracted and simpler lifestyle. As with suburban living, a less transient "small town" environment promotes closer long-term ties to friends and neighbors, active interest and engagement in community affairs and activities structured and centered upon children and extended families.[200]

Assuming now that these broad generalizations are more true than flawed, how would variously prophesied smart tech promises and problems disproportionately impact each relative to the others?

Would comprehensively wired-together smart city efficiencies, on-call driverless cars and ubiquitous anti-crime surveillance monitoring be readily acceptable in more suburban and outlying locations?

Would large corporate offices downsize their metropolitan facilities and operations due to less reliance on in-house personnel and a shift to telecommuting employees and remote as-needed expertise, therein reducing their host city's real estate, business and employee tax revenues to finance those smart city updates?

Would suburbs and rural communities, in balance, gain resident populations, business start-ups and services, infrastructure investment capital and tax revenues, and private and public citizen employment opportunities inherited from corporate decentralization and business exodus from high urban taxation, regulation and physical and social deterioration?

Would the more metro-independent suburbs and rural communities be less reliant upon, and thereby less negatively impacted by disruptive AI and automation developments due to availability of less expensive human labor and increasingly person-centric markets such as healthcare and community-based education programs?

How will each type of community redefine and optimize its most attractive features in order to adapt, where will the resources come from and what is the federal government's role and responsibility in supporting these transitions?

And the biggest question of all: how would variously proposed remedies to unemployment, workforce realignments and "income equity" stratification impact America's overall long-term socio-economic future?

So first, let's just imagine that these urban trends, aided and abetted by technology, advance as suggested. In this scenario, major businesses will decentralize away from costly big city real estate and taxes. Telecommuting and remote consulting will extend greater elective freedom for family income providers to reside in more affordable and safer places to raise children. Growing suburban and rural human service needs such as medicine for aging populations will be augmented by telemedicine, clinics and a larger abundance of health and assisted care workers. Less transient, more neighborhood-friendly communities promote civic engagement and pride.

And yes, private automobile and truck ownership remains both necessary and cherished as an individual freedom.

Yet let's also remember that cities offer important lifestyle and cultural attractions that will hold enduring treasures as well. Each is uniquely special, culturally diverse, historically significant, energetically stimulating and organically evolving and enriching. Perhaps noisy and congested, their residents and endless streams of visitors encounter and celebrate wonderful architectures and impressive centers of commerce, marvelous world-class museums and live theater, thriving hustle and bustle of activity and a vivid, multi-hued spectrum of humanity and enterprise.

In any case, any technology-driven flight from large cities won't happen either quickly or completely. Nor will metropolitan transformations to bright, visionary sci-fi-style centers with all-things automated and modern.

No, it won't be like that at all. Those new smart city utopian-marketed infrastructures will be built upon and around the bones and hulking shells of aging cores. Many improvements are likely to occur to the detriment of older traditions and localized lifestyles that will be left behind and displaced in exchange for technological efficiencies and conveniences.

John Steinbeck painted a grim picture of accelerating outcomes in his 1960 book *Travels with Charley: In Search of America:*

> *When a city begins to grow and spread outward from the edges, the center which was once its glory...goes into a period of desolation inhabited at night by the vague ruins of men. The lotus eaters who struggle daily toward unconsciousness by the way of raw alcohol. Nearly every city I know has such a dying mother of violence and despair where at night the brightness of the street lamps is sucked away and policemen walk in pairs.*

Perhaps more optimistically, Steinbeck then adds:

> *And then one day perhaps the city returns and rips*

out the sore and builds a monument to its past.

The challenge at hand, however, is for human society to determine in advance what cultural and economic monuments to achievements and failures will ultimately represent America's future.

Capitalism Obituaries are Premature

Silicon Valley gurus predict that governments will ultimately need to pay people whose livelihoods have been displaced by AI a universal basic income, freeing them to pursue their dreams unburdened by the need to earn a living. In another scenario, it will generate staggering wealth inequities, chaos and failed capitalist states across the globe.

Technology writer David Rotman warns in the *MIT Technology Review* that the relentless pace of automation "will not be the normal churn of capitalism's creative destruction, a process that inevitably arrives at a new equilibrium of more jobs, higher wages and better quality of life for all." [201]

Rotman ominously forecasts that Steinbeck's dire view of social despair will become widespread throughout the country and beyond. He forecasts that the coming AI economy will cause breakdowns in many of the free market's self-correcting mechanisms leading to a new 21st century caste system—one hopelessly split into a plutocratic AI elite and powerless struggling masses. He writes:

> *If we allow AI economics to run their natural course, the geopolitical tumult of recent years will look like child's play. I fear that [job losses] will lead to a crushing feeling of futility and obsolescence. At worst, it will lead people to question their own worth and what it means to be human.*

So what can be done?

Kai-Fu Lee, who incidentally is the former president of Google China, says that the grim view of AI impacts upon job losses shared by many Silicon Valley technologists has sent them casting about for quick-fix solutions. As the architects and profiteers of the AI age, he attributes their concern to a mix of genuine social responsibility and fear of being targeted when the pitchforks come out. The answer the techno-elites have seized upon is as old as the roots of socialism—a guaranteed government-provided Universal Basic Income (UBI) for everyone.

Writing in the *Wall Street Journal*, Lee observes:

> *(This is) what Silicon Valley entrepreneurs love: an elegant technical solution to tangled social problems. UBI can be the magic wand that lets them wish away the messy complexities of human psychology and get back to building technologies that 'make the world a better place,' while making them rich. It's an approach that maps well onto how they tend to view society: as a collection of 'users' rather than as citizens, customers and human beings.*[202]

Nevertheless, Lee doesn't actually rule out his own advocacy for "some form of guaranteed income," but believes that such support should not be viewed as an endgame solution. In doing so, Lee says, we will miss out on the opportunity presented by this transformative technology. He writes:

> *Instead of simply falling back on an economic painkiller like universal basic income, we should use the economic bounty generated by AI to double down on what separates us from machines: human empathy and love.*

Dr. Lee urges that the AI revolution will require comprehensive rethinking about how humans relate to work from all corners of society. He writes:

> *In the private sector, instead of simply viewing AI as a means for cost-cutting through automation, businesses can create jobs by seeking out symbiosis between AI optimizations and the human touch. This will be especially powerful in areas such as health care and education, where AI can produce crucial insights but only humans can deliver them with care and compassion.*[203]

Whereas it's hard to argue against the idea that private sector businesses should work to apply smart tech compassionately with human employee and client interests in mind, Lee then goes on to suggest that government will then still need to jump in and do what it does best—namely to provide subsidized jobs for those wise, socially benevolent bureaucrats deem most "beneficially" worthy.

Lee proposes:

> *At the center of this vision, I would suggest, there needs to be what I call the Social Investment Stipend, a respectable government salary for those who devote their time to three categories of activities: care work, community service and education. These activities would form the pillars of a new social contract, rewarding socially beneficial activities just as we now reward economically productive activities. The idea is simple: to inject more ambition, pride and dignity into work focused on enhancing our communities.*[204]

According to Lee's hypothesis, requiring some social

contribution in order to receive the stipend would foster a public philosophy far different from laissez-faire individualism of universal basic income. I fail to see a distinction.

Lee's proposed social public works program would compete with, not invigorate, private sector job creation and economic growth initiatives by providing de facto government-sponsored bribes for those who support politically-favored flavor of the month agendas. As for fostering a public philosophy different from the laissez-faire individualism of universal basic income, it only substitutes prescriptive public servitude.

Economic historian Joel Mokyr at Northwestern University says that while societies that have experienced and survived radical past technological transitions such as the Industrial Revolution in the late 18th century, this one is occurring much faster and "more intensive." [205]

Mokyr describes himself as "less pessimistic" than others about whether AI will create plenty of jobs and opportunities to make up for those lost, but cites another troubling quandary:

> *There is no question that in the modern capitalist system your occupation is your identity," and the pain and humiliation felt by those whose jobs have been replaced by automation is 'clearly a major issue.'*

He adds,

> *I don't see an easy way of solving it. It's an inevitable consequence of technological progress.*

Even Kai-Fu Lee points out that many techno-optimists will argue that productivity gains from new technology almost always create more jobs and prosperity than before, although all opportunities aren't created equal. Some changes replace one kind of labor (the calculator), while other inventions—the cotton gin for example—can disrupt entire industries.

AI won't merely effect a single industry, but rather will drive changes across hundreds of them. This has happened only three times in the past three centuries, with the steam engine, electrification and now information technology.[206]

I'll add here that the steam engine and electrification which led to the Industrial Revolution created more jobs than they destroyed, in part by breaking down the work of a single craftsman into simpler tasks done by dozens of factory workers. Whereas the craftsmen may have earned proportionately larger incomes, the factories produced more jobs and more affordable products that benefitted more people.

Industries that grew out of the Industrial Revolution included new businesses to invent, produce and sell new products through competitive markets that drove more innovation, more businesses and more workplace and consumer options.

Those new small and startup businesses will benefit hugely from the Information Revolution as well. Remember that the first "supercomputer," the Control Data Corporation (CDC) 6600 released in 1964 cost $8 million—around $60 million in today's money—with far less processing capacity than in current smart phones. But by comparison, CDC 600 was still up to 10 times faster at the time than the IBM 7030 "Stretch," which took up 2,000 square feet of space and cost $13 million. Only large corporations could afford such technological advantages.

True, and as always, these new personal job and income opportunities are not all created equal. There is really nothing new about "income inequality," and certainly not under socialist regimes. Nor should we expect this to be the case with the Information Revolution.

One major difference, perhaps, is that the Information Revolution will disrupt disproportionately higher-income white-collar professionals as well as traditionally lower-paid blue-collar workers.

Dr. Lee acknowledges that jobs that are asocial and repetitive are most vulnerable to displacement by AI and automation, along with many semi-social core tasks associated with medical doctors and other well-compensated professionals. The safest near-term jobs are those beyond the reach of AI's capabilities in terms of creativity, strategy and sociability, from social workers to CEOs.

Notably, some of the jobs that Dr. Lee predicts will remain most insulated from AI disruption appear to fall in the same socially interactive categories such as home care workers, and presumably teachers as well, which he slated for generous taxpayer-funded government subsidy stipends. It's probably not realistic to expect that such charity will be freely donated by big data tech largess.

According to the McKinsey & Company Global Institute, in 2008, Microsoft was the only tech company that made it into the top ten largest companies globally. Apple came in next at 39, and Google at 51. By 2018, the top five spots were claimed by the top five tech giants, both in the United States and globally.

At the same time, however, tech giants create relatively few jobs competitively to their hold on the market. In 1990, Detroit's three largest companies were valued at $65 billion with 1.2 million workers. In 2016, Silicon Valley's three largest companies were valued at $1.5 trillion, but with only 190,000 workers.[207]

As for vaunted big tech interests in those smart cities with driverless cars and trucks, David Rotman estimates in MIT Technology Review that automated vehicles could threaten or alter 2.2 million to 3.1 million existing jobs. This includes 1.7 million jobs driving tractor-trailers, the heavy rigs that dominate highways.[208]

The inescapable reality here is that no one can predict what new industries and jobs will be created, transformed or eliminated by AI-driven influences. Another reality is that socialist trending "government knows best solutions" to ameliorate disproportionately difficult impacts upon some groups relative to others are costly charities which will only impoverish human initiative and progress.

According to a report titled *Artificial Intelligence, Automation, and the Economy* report released by top economic advisors of the outgoing Obama administration just six weeks following President Trump' election said:

> The problem is that the United States has been particularly bad over the last few decades at helping people who've lost out during periods of technological change. Their social, educational, and financial problems have been largely ignored, at least by the federal government.[209]

In truth, the American people have not overlooked the needs of its poorest citizens. Subsequent to the mid-1960s, federal and state subsidies for low-income families have increased steadily, yet with no proportionate reduction in their numbers relative to the general population.[210]

For clarification, it's important to understand that the U.S. Census Bureau counts as poor all people in families with incomes lower than the established income thresholds set in 1963 adjusted for inflation for their respective family size and composition. The bureau reported in 2016 that some 12.7 percent of Americans lived in poverty, a rate that has remained virtually unchanged over the past 50 years.

Meanwhile, the total inflation-adjusted government-transfer payments to low-income families in real dollars increased from an average of $3,070 per person in 1965, to $34,093 in 2016.[211]

Even those numbers significantly understate other

payments to low-income families. The bureau measures poverty by what it calls "money income," which includes some transfer payments such as Social Security and unemployment insurance. Yet it excludes food stamps, Medicaid (the portion of Medicare going to low-income families), Children's health Insurance, the refundable portion of the earned-income tax credit, at least 87 other means-tested federal payments to individuals and most means-tested state payments.

Medicare and Social Security provide large subsidies to low-income retirees compared with what they pay into taxes. The lowest quintile of earners can receive as much as ten times the lifetime benefits received by the highest quintile earners, and three times as much as the middle quintile.

If government counted these missing $1.5 trillion in annual transfer payments, the poverty rate would be less than 3 percent. That rate would fall even further if it accounted for transfers within families of some $500 billion of private charitable giving.

None of this has worked out as planned when President Lyndon Johnson declared in 1964:

> *The War on Poverty is not a struggle simply to support people. It is an effort to allow them to develop and use their capacities.*

That goal was to make America's poor more self-sufficient and to bring them into the mainstream of the economy—not to increase dependency and largely sever the bottom fifth of earners from the rewards and responsibilities of work.

In 1965, prior to the time that program was funded, more families in the lowest income quintile with a head of household of prime working age who had at least one person employed outnumbered those that didn't. A broadly evident work ethic revealed that the lowest quintile of families with working-age heads and no one working was only 5.4 percent higher than those in the middle quintile.

By 1975, the lower fifth of families had 24.8 percent more families with a prime-work age head and no one working than did the middle-income sector. By 2015, this differential had risen to 37.2 percent.

As President Franklin Roosevelt providently observed in 1945:

> *The lessons of history show conclusively that continued dependency upon relief induces a spiritual and moral disintegration fundamentally destructive to the national fiber."*

Some Grossly Speculative Scenarios

Fast forwarding now to some lessons and projections that might be drawn from all previous narrative, what futures—near term and beyond—might await?

Will job losses resulting from ever more advanced and ubiquitous AI-driven data processors and automated systems drive America to embrace big tech promises of socialist salvation?

Will Americans ultimately surrender personal privacy and independent free will to mega-powerful data mining and social media enterprises for "the public good"?

Will ever-expanding networks of Internet-connected high tech services and conveniences detrimentally transform value and lifestyle expectations?

Will AI-driven job and career displacements, realignments and widening income disparities lead to regional conflicts and ideological class warfare?

Will ever-smarter self-teaching devices ultimately render their human inventors—and therein, prevailing concepts of general intelligence—obsolete altogether?

No experts exist who can offer anything other than wildly speculative conclusions regarding any of these very general questions. Answers to each, along with myriad others, are in turn interconnected and interdependent with countless layers of unknowable issues.

Economist and technology writer George Gilder counsels us to recognize that surprise is the requisite determinate of invention, just as unpredictability of success fuels the driving competitive force of capitalism. Nowhere are the elements of unpredictably and surprise more salient than with regard to information technology development, applications and social impacts.

Having offered this as a de facto disclaimer of attaching any credibility of what will immediately follow, I'll just go ahead and propose one rather dismal scenario anyway. Let's imagine, for example, the following:

Smart City Benefits Prove Costly

Empowered by big tech largess, unchallenged political influence and persuasive promises of benefits to efficiency and public safety, centrally monitored and controlled Internetworks erase virtually all boundaries between personal and public life decision entitlements. Under such circumstances, those social control forces purporting to serve "a greater public good" will inevitably win out over individualist free choice advocate hold-outs.

Metropolitan private automobile ownership will become taxed out of existence through parking and street-use fees, and personal cars will ultimately be banned altogether. Private vehicles will initially be required to be electric plug-in varieties premised upon urban pollution and CO_2 "climate protection" arguments, followed by restrictions favoring or mandating use of on-demand taxi services (driverless or not) to reduce traffic congestion and non-use parking requirements. Private taxi services will soon become replaced by quasi-public operations which are later integrated entirely under metropolitan transit system management authorities.

Massive municipal government purchases of automated on-demand vehicles will afford high leveraging over the traditional private automotive market-driven industry by big tech and powerful political allies. This influence will extend to large pricing penalties imposed upon production and sales of internal combustion engine vehicles, which in turn, will cause electricity prices to soar.

In response, expect to see energy rationing based upon unit allowances allocated on the basis of political social justice equity merit determinations. Wired-together power distribution networks connecting home appliances and centrally monitored and algorithm controlled temperature thermostats will prevent unfair cheating under penalty of fines and risks of service disconnection.

The same rationing system may apply to energy units used in travel. Recharging of private plug-in cars will automatically draw upon allowances, but on-demand taxi services will be registered to personal energy service accounts as well. All travel will be visually monitored and recorded from inside and outside the vehicles as a public safety precaution.

AI and Internet-driven developments will impact city economies in two fundamental ways. First, telecommuting and increased use of Internet-wired-together expertise will enable large metropolitan-based companies to cut back on expensive real estate holdings, centralized staff accommodations and constantly escalating tax burdens to compensate for lost municipal revenues caused by others that follow this same trend. Second, decentralization of businesses will negatively impact the human service industry that supports these employment centers, further reducing municipal revenues...including those required to pay for added costs of fulfilling smart city promises.

Whereas some special urban attractions will thrive, specialty shopping, museum and theater districts, and global commerce and conference centers requiring proximity to major airports, for example, will fare far less fortunately. Increased living costs will disproportionately burden the poorest residents, public welfare demands will exceed taxing capacities, tenement warehousing of indigent families and individuals will exacerbate social distress and crime and reliance upon electronic surveillance to maintain public order will advance police state justifications.

Capitalism Flees to the Countryside

Revenue-depleted metropolitan centers will attempt to expand their revenue bases and taxing authority through "public good" eminent domain land grabs and annexation of more prosperous adjacent communities to commandeer outside subsidies needed to offset growing infrastructure and utility deficits. Such private property and taxation encroachments will not be welcomed by most of those whose personal liberties and fruits of labor are targeted for social justice sacrifice.

While AI-driven technologies, including automation, will affect all job sectors, community settings and geographic regions, overall they will impact outlying suburban and rural areas the least; along with also affording certain inherent advantages. Among these, decentralized work and enterprise opportunities will enable more and more people to live where they wish: family-friendly communities that are affordable, safe places with good schools to raise their children and natural and social surroundings that appeal to individual lifestyle priorities.

Low costs and expanding capabilities of Information technology enable and encourage small businesses and start-ups along with larger community-based enterprises and institutions to serve ever broadening service markets which large strategically-located corporations previously monopolized. Included are online blog writers and publications that are supplanting large media domination; online community-based educational courses and programs that will supplant and enhance services traditionally provided by expensive brick and mortar campuses; and smaller regionally distributed medical clinics that apply huge online datasets and remote AI telemedicine applications to broader populations.

Suburban and rural areas will provide abundant and moderate cost human labor resources for businesses and institutions that prioritize interactive personal attention and care. This will continue to include those who will attend to health and home care needs of an increasingly older population that is resulting, in part, from medical technology advancements. Human innovative capabilities, problem solving skills and versatility will also be required to support the operations and maintenance of those automated systems of production and distribution that businesses and industries will always require.

And don't expect a robot to come to your home to fix an urgent plumbing leak, to build a new porch on the back of your home or to repair that personal car that has yet to be outlawed.

A Luddite-Lapdog Revolution:

Deeping ideological divisions between predominately metropolitan and countryside populations will have profound national, state and local political consequences. Regions of the United States with largest cities will be disproportionately impacted by AI-driven demographic shifts.

Offers of guaranteed incomes to those whose jobs are displaced along with failed promises that government-distributed economic proceeds of AI and automation will liberate humans from labor will not sit well with country-wise Luddites who recognize these socialist agenda scams as exactly what they are.

Luddite individualists and communities won't willingly trade in their freedom and convenience to travel the countryside wherever and whenever they wish in private vehicles in exchange for public transit tokens to destinations of dwindling or remote interest. They will bristle at lapdog attempts to turn virtually all roadways into tollways in order to subsidize costly public transit networks that provide inefficient and poor service in outlying suburban and rural areas. They will continue to drive their children to school, load groceries in their personal cars and carry building supplies and other cargo in their vans and pickup trucks.

The Luddites will favor neighbor recognition over electronic facial recognition in a community culture where people look after one another out of view of privacy-intrusive surveillance cameras. The Luddites will honor the importance of civic participation over civil protest, and community values that emphasize the importance of earned work ethics above income-equality entitlement charities.

Luddite enterprise will prosper at the expense of lethargic lapdog losses. Industries and businesses will follow work-motivated labor workforces to more affordable and safer countryside settings. Online entrepreneurship will flourish, including small work-from-home startups and retirement consulting. Online shopping and spacious land will attract more and more retail storage and outlet centers to serve growing populations. The demand for construction and maintenance services will grow, along with opportunities for a host of other skilled trades and professions.

Existing ideological and political tensions between lapdogs and Luddites will intensify. Lapdog anger will be inflamed over convictions that the Luddites are resisting the blessings and costs of technical nirvana promised by smart city efficiencies and comforts. The Luddites will resent and rebel against the lapdog

Reinventing Ourselves

culture of passivity, indolence and acquiescence to technology barons bearing costly socialist gifts they must pay for.

Which side will ultimately win and lose out in this contentious scenario? Very likely there will be some wins and losses both ways. Although I am admittedly rooting for the Luddites, perhaps there is still some final hope for the lapdogs as well. Maybe they will finally prove to be smarter than those hapless frogs sitting complacently in a pan while the water gradually heats to a boil after all.

Stephen Hawking has warned that artificial intelligent machines could wind up killing us all by themselves because they become too clever. Responding to a questioner during an *Ask Me Anything* session on Reddit, he replied:

> One can imagine such technology outsmarting financial markets, out-inventing human researchers, out-manipulating human leaders, and developing weapons we cannot even understand. Whereas the short-term impact of AI depends on who controls it, the long-term impact depends on whether it can be controlled at all.[212]

Finally, let's recognize that all heaven-or-hell scenarios such as any of these are like winning the Powerball jackpot...extremely unlikely.

Can we expect that future generations will witness angry protesters shouting "Hey hey, ho ho, AI overlords must go!"? Probably not.

Will smart tech ultimately outsmart humanity? No. Humanity can only surrender the dominion over its genius and the triumphant mastery over its inventions.

Let's give our own marvelous intelligence, creativity and judgment more credit than that.

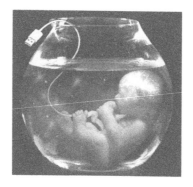

Endnotes

[1] "Where machines could replace humans—and where they can't (yet)," Michael Chui, James Manyika, and Mehdi Miremadi, McKinsey Quarterly, July 2016.
[2] "The history of computing is both evolution and revolution," Justin Zobel, The Conversation.com, May 31, 2016.
[3] Ibid.
[4] "Ken Olsen: Did Digital founder Ken Olsen say there was 'no reason for any individual to have a computer in his home"?, Snopes.com, https://www.snopes.com/fact-check/ken-olsen/.
[5] "Why GANs give artificial intelligence wonderful (and scary) capabilities," Gabriel Sidhom, August 23, 2017, Orangesv.com.
[6] Ibid.
[7] "Rise of the Robots," Stephen Moore, Newsmax, October 2017.
[8] "AI as Servants or Spies?" John Edwards, Newsmax, October 2017.
[9] "Rise of the Robots," Stephen Moore, Newsmax, October 2017.
[10] "Quantum Computing May Outsmart All Cyber Defenses," Larry Bell, Newsmax, November 12, 2017.
[11] "10 Breakthrough Technologies," MIT Technology Review, 2018.
[12] "Post-quantum cryptology—dealing with the fallout of physics success," Daniel J. Bernstein and Tanja Lange, National Science Foundation/European Commission, April 9, 2017.
[13] "We'll Need Bigger Brains Keeping Up With the Machines," Christof Koch, Wall Street Journal Review Section, October 28-29,

2017.

[14] Ibid.

[15] "How cloud computing is changing the world…without you knowing," Joe Baguley, September 24, 2013, The Guardian.com, Media network blog.

[16] "Cloud Computing is Crucial To The Future Of Our Societies—Here's Why," Joy Tan, Feb. 25, 2018, Forbes.com.

[17] "Cloud Computing; Current and Future Impact on Organizations," Yiyun Zhu, March 20, 2017, Western Oregon University, Department of Computer Science student thesis paper, Digital Commons@WOU.

[18] "Cloud Computing is Crucial To The Future Of Our Societies—Here's Why," Joy Tan, Feb. 25, 2018, Forbes.com.

[19] Bernard Golden's 5 predictions for cloud computing in 2018, Bernard Golden, blog, https://cloud.withgoogle.com/build/leadership/five-predictions-cloud-computing-2018/.

[20] "How the Internet Has changed Everyday Life," Taryn Dentzel, Openmind.com.

[21] Ibid.

[22] "What Is the Internet Doing to Relationships?" John B. Horrigan, Barry Wellman and Jerry Boase, Pew Research Center, Internet and Technology, January 25, 2006 http://www.pewinternet.org/2006/01/25/what-is-the-internet-doing-to-relationships/.

[23] "The Effect OF Technology On Relationships," Alex Lickerman, M.D., June 8, 2010, Psychology today.com, https://www.psychologytoday.com/us/blog/happiness-in-world/201006/the-effect-technology-relationships.

[24] "How the Internet Has changed Everyday Life", Zaryn Dentzel, Openmind.com.

[25] "Couples, the Internet, and Social Media, How American couples use digital technology to manage life, logistics, and emotional intimacy within their relationships", Amanda Lenhart and Maeve Duggan, February 11, 2014, Pew Research Center, Internet & Technology http://www.pewinternet.org/2014/02/11/couples-the-internet-

and-social-media/.

[26] "Cyberbullying websites should be boycotted, says Cameron; prime Minister calls for website operators to 'step up to the plate', following death of 14-yeat-old Hannah Smith," Alexandra Topping, Ellen Coyne and agencies, August 8, 2013, The Guardian.

[27] "Yahoo, Bucking Industry, Scans Emails for Data to Sell," Douglas MacMillan, Sarah Krouse and Keach Hagey, August 8, 2018, Wall Street Journal.

[28] "Yahoo, Bucking Industry, Scans Emails for Data to Sell," Douglas MacMillan, Sarah Krouse and Keach Hagey, August 8, 2018, Wall Street Journal.

[29] "AI as Servants or Spies," John Edwards, Newsmax, October 2017.

[30] Ibid.

[31] "It's not paranoia, your phone really IS listening to EVERYTHING you say and using your private conversations to target ads, security expert warns," Harry Pettit, June 7, 2018, Daily Mail online.

[32] "This IS Your Brain On the Internet: Why the world seems worse than it is; Our primitive biases and fears are magnified by online algorithms," Christopher Mims, the Wall Street Journal, September 1-2, 2018.

[33] "When a smart home becomes a trap," Robby Berman, July 1, 2018, New York Times.

[34] "How is the Internet changing the way we behave?", Mary Aiken, March 27, 2017, Science Focus.com.

[35] "Computers 'changing our values, language, culture,'" Mary K. Pratt, June 3, 2009, ITBusiness.ca.

[36] "Information Technology and Moral Values," Stanford Encyclopedia of Philosophy, First published, June 12, 2012.

[37] "How Has Social Media Changed Us?", Carrie Kerpen, April 21, 2016, Forbes.com.

[38] Ibid.

[39] Ibid.

[40] "How is the Internet changing the way we behave?", Mary Aiken, March 27, 2017, Science Focus.com.

[41] "Tired of Swiping Right, Some Singles Try Slow Dating," Kari Paul, August 28, 2018, Wall Street Journal.

[42] "What Will Our Society Look Like When Artificial Intelligence Is Everywhere? Here are five scenarios from our future dominated by AI," Stephan Talty, Smithsonian Magazine.
[43] "Slow Dating Instead of Swiping Right," Karl Paul, August 29, 2018, Wall Street Journal.
[44] "Labor of Love: The Invention of Dating," Moira Weigel, 2016, Ferrar, Straus and Giroux.
[45] "How the Internet Has changed Everyday Life," Zaryn Dentzel, Openmind.com.
[46] Is Google Making Us Stupid?", Nicholas Carr, July/August, 2008, The Atlantic.
[47] "How Has Google Affected The Way Students Learn?", PBS, KQED.org Mindshift.
[48] Is Google Making Us Stupid?", Nicholas Carr, July/August, The Atlantic.
[49] Ibid.
[50] Ibid.
[51] "All Eyes on Google," March 28, 2004, Newsweek, https://www.newsweek.com/all-eyes-google-124041.
[52] Is Google Making Us Stupid?", Nicholas Carr, July/August, The Atlantic
[53] Ibid.
[54] Ibid.
[55] "20 Industries Threatened by Tech Disruption," Adam Hayes, Investopedia, Feb. 6, 2015.
[56] "Learning Representations for Multimodal Data with Deep Belief Nets", Nitish Srivastava and Ruslan Salakhutdinov, Presented at the ICML Representation Learning Workshop, Edinburgh, Scotland, UK 2012, http://www.cs.toronto.edu/~nitish/icml2012/paper.pdf.
[57] "Should Artificial Intelligence Copy the Brain?", Christopher Mims, Wall Street Journal, August 4-5, 2018
[58] Ibid.
[59] Ibid.
[60] "Harnessing automation for a future that works," Lames Manyika, Michael Chui, Mehdi Miremadi, Jacques Bughin, Katy George, Paul Willmott, and Martin Dewhurst, McKinsey & Company/ McKinsey

Global Institute, January, 2017.
[61] "Where Do Humans Outperform AI?", Kevin McCaney, Governmentciomedia.com, April 4, 2018.
[62] Ibid.
[63] "Will AI Best All Humans Tasks by 2060? Experts Say Not So Fast," Alisa Vakludes Whyte, Huffington Post, June 25, 2017,
[64] Ibid.
[65] "Industrial robots will replace manufacturing jobs—and that's a good thing," Matthew Randall, TechCrunch.com, October 9, 2016.
[66] Jobs lost, jobs gained: What the future of work will mean for jobs, skills and wages, McKinsey & Company, James Manyika, Susan Lund, Michael Chui, Jacques Bughin, Jonathan Woetzel, Paul Batra, Ryan Ko, and Saurabh Sanghvi, November, 2017.
[67] Ibid.
[68] "Every study we could find on what automation will do to jobs, in one chart," Erin Winick, MIT Technology Review.com, January 25, 2018.
[69] "20 Industries Threatened by Tech Disruption," Adam Hayes, Investopedia, Feb. 6, 2015.
[70] "20 Industries Threatened by Tech Disruption," Adam Hayes, Investopedia, Feb. 6, 2015.
[71] "11 Industries Being Disrupted By AI," David Roe, CMS Wire.com, April 27, 2018.
[72] "5 Fast-growing Technology Trends for 2017," Cynthia Harvey, Cloud Infrastructure, McAfee, Jan. 11, 2017.
[73] "Will Robots Take Our Children's Jobs?", Alex Williams, New York Times, December 11, 2017.
[74] "Where machines could replace humans—and where they can't (yet)," Michal Chui, James Manyika, and Mehdi Miremadi, McKinsey Quarterly, McKinsey & Company, July 20, 2016.
[75] "Jobs lost, jobs gained: What the future of work will mean for jobs, skills and wages," McKinsey & Company, James Manyika, Susan Lund, Michael Chui, Jacques Bughin, Jonathan Woetzel, Paul Batra, Ryan Ko, and Saurabh Sanghvi, November, 2017.
[76] "Where machines could replace humans—and where they can't (yet)," Michal Chui, James Manyika, and Mehdi Miremadi, McKinsey

Quarterly, McKinsey & Company, July 20, 2016.

[77] "Jobs lost, jobs gained: What the future of work will mean for jobs, skills and wages," McKinsey & Company, James Manyika, Susan Lund, Michael Chui, Jacques Bughin, Jonathan Woetzel, Paul Batra, Ryan Ko, and Saurabh Sanghvi, November, 2017.

[78] "The 3-D Printing Revolution," Richard D'Aveni, Harvard Business Review, May 2015 Issue.

[79] "How AI Is Transforming The Future Of Healthcare," Gunjan Bhardwaj, Forbes.com, Jan. 30, 2018.

[80] "Artificial Intelligence in Healthcare: Separating Reality From Hype," Robert Pearl, Forbes.com, March 13, 2018.

[81] "Tomorrow's Lawyers: An Introduction to Your Future," Richard Susskind, Oxford University Press, 1999, ISBN 978019968069.

[82] "Legal Technology: Artificial Intelligence and the Future of Law Practice," Mark McKamey, Appeal Law Journal—Vol. 22, Citation (2017) Appeal 45.

[83] "Why Robots Will Not Take Over Human Jobs," Andrew Arnold, Forbes.com, March 27, 2018.

[84] "Jobs lost, jobs gained: What the future of work will mean for jobs, skills and wages," McKinsey & Company, James Manyika, Susan Lund, Michael Chui, Jacques Bughin, Jonathan Woetzel, Paul Batra, Ryan Ko, and Saurabh Sanghvi, November, 2017.

[85] "Will Jobs Exist in 2050?", The Guardian, Feb. 21, 2017.

[86] "Out of the Office: More People Are Working Remotely, Survey Finds Out of the Office: More People Are Working Remotely, Survey Finds," Niraj Chokshi, February 15, 2017, the New York Times.com.

[87] "Why Remote Work Thrives in Some Companies and Fails in Others," Sean Graber, March 20, 2015, Harvard Business Review.

[88] "How Remote Work Is Changing And What It Means For Your Future," William Arruda, February 16, 2017, Forbes.com.

[89] "The Future of Home Business Technology," Tiffany S. Williams, December 11, 2017, Forbes.com.

[90] "Benefits of Telecommuting For The Future of Work," Andrea Loubier, July 20, 2017, Forbes.com.

[91] "The truth about smart cities: 'In the end, they will destroy democracy,'" Steven Poole, December 17, 2014, The Guardian.

[92] "Stop Saying 'Smart Cities,'" Bruce Sterling, February 12, 2018, The Atlantic.com, Technology.
[93] "The smart entrepreneurial city: Dholera and 100 other utopias in India," S. Marvin, A. Luque-Ayala, and C McFarlane (Eds.), 2015, Smart urbanism: Utopian vision or false dawn?, Rutledge.
[94] "The truth about smart cities: 'In the end, they will destroy democracy,'" Steven Poole, December 17, 2014, The Guardian.
[95] Ibid.
[96] "Smart Cities and the Idea of Smartness in Urban Development—a Critical Review," Milan Husar, Vladimir Oudrejieka and Sila Ceren Varis, 2017, IOP Conference Series; Materials Science and Engineering 245.
[97] Ibid.
[98] "The imaginary real world of cybercities," M.C. Boyer, 1992, Assemblage.
[99] "A model Korean ubiquitous eco-city? The politics of making Songdo," S.T. Shwayri, 2013, Journal of Urban Technology.
[100] "The truth about smart cities: 'In the end, they will destroy democracy,'" Steven Poole, December 17, 2014, The Guardian.
[101] "Who Can You Trust? How Technology Brought Us Together and Why It Might Drive Us Apart," Rachel Botsman, (2017) Penguin Books.
[102] Ibid.
[103] "Baidu, Alibaba, and Tencent: The Rise of China's Rech Giants," Peter Diamandis, August 17, 2028, SingularityHub.com.
[104] "Who Can You Trust? How Technology Brought Us Together and Why It Might Drive Us Apart," Rachel Botsman, (2017) Penguin Books.
[105] Ibid.
[106] Ibid.
[107] "Baidu, Alibaba, and Tencent: The Rise of China's Rech Giants," Peter Diamandis, August 17, 2028, SingularityHub.com.
[108] "Who Can You Trust? How Technology Brought Us Together and Why It Might Drive Us Apart," Rachel Botsman, (2017) Penguin Books.
[109] Ibid.

110 "China is building a vast civilian surveillance network—here are 10 ways it could be feeding its creepy 'social credit system,'" Alexandra Ma, April 29, 2018, Business Insider.com.
111 "China's Surveillance State Should Scare everyone," Anna Mitchell and Larry Diamond, February 2, 2018, The Atlantic.
112 "Self-driving cars will set off an economic and cultural earthquake," Eric Risberg, May 11, 2018, The Global Mail.com.
113 "How Driverless Cars Will Change the Feel of Cities," Ian Bogoost, November 15, 2017, The Atlantic.com.
114 Ibid.
115 "The Jeep hack was only the beginning of smart car breaches," Cadie Thompson, July 22, 2015, Business Insider.com.
116 "Hackers Might be Able to Take Control of your Smart Car," Michael Aechambault, March 9, 2018, PSafe.com.
117 "How Driverless Cars Will Change the Feel of Cities," Ian Bogoost, November 15, 2017, The Atlantic.com.
118 "Models Will Run the World," Steven A. Cohen and Matthew W. Granade, Wall Street Journal, August 20, 2018.
119 "Trading Places," Liz Hoffman and Telis Demos, August 18, 2018, Wall Street Journal.
120 "Commentary: The biggest risk to the market? The market itself," Jamie McGeever, February 6, 2018, UK.Reuters.com.
121 "The stock market is controlled by algorithms that are fighting with each other," Annalee Newitz, May 11, 2011, i09.gizmodo.com.
122 "Code: And Other Laws of Cyberspace, Version 2.0/Edition 2," Lawrence Lessig, December 28, 2006, Basic Books.
123 "They Are Watching You," Robert Draper, February 2018, National Geographic.
124 "Welcome to the Quiet Skies," Jana Winter, July 28, 2018, the Boston Globe.
125 "They Are Watching You," Robert Draper, February 2018, National Geographic.
126 Ibid.
127 Ibid.
128 Ibid.
129 "We let technology into our lives. And now it's starting to control

us," Rachel Holmes, November 28, 2016, The Guardian.com.
130 Google Eyed Check on Travel Ban," John D. McKinnon and Douglas MacMillan, September 21, 2018, the Wall Street Journal.
131 Ibid.
132 "Yes, Biased Social Media Platforms Can Censor," Larry Bell, September 10, 2018, Newsmax.com.
133 Ibid.
134 "Google Eyed Check on Travel Ban," John D. McKinnon and Douglas MacMillan, September 21, 2018, the Wall Street Journal.
135 Ibid.
136 "Ex-Google engineer who worked on China search engine calls out 'wrong,'" Daniel Howley, September 19, 2018, Yahoo Finance.
137 "We let technology into our lives. And now it's starting to control us," Rachel Holmes, November 28, 2016, The Guardian.com.
138 Ibid.
139 "The Master Algorithm: How the Quest for the Ultimate Learning Machine Will Remake Our World," Pedro Domingos, September 2015, Basic Books.
140 "Peter Domingos on the Arms Race in Artificial Intelligence," Interview by Christopher Sheuermann and Bernhard Zand, August 16, 2018, Spiegel Online.
141 Information Technology and Moral Values," Stanford Encyclopedia of Philosophy, June 12, 2012.
142 "Toward an Approach to Privacy in Public: Challenges of Information Technology," Ethics and Behavior, 7(3):207-219 Hellen Nissenbaum, 1997.
143 "Toward an Approach to Privacy in Public: Challenges of Information Technology," Ethics and Behavior, 7(3):207-219 Hellen Nissenbaum, 1997.
144 Ibid.
145 Beijing's Big Brother Tech Needs African Faces," Amy Hawkins, July 24, 2018, Foreign Policy.com,
146 Ibid.
147 "The Rise of the Machines," Richard Dooling, October 11, 2008, The New York Times.com.
148 Ibid.

[149] Ibid.
[150] "Sapiens: A Brief History of Humankind," Yuval Noah Harari, 2015, HarperCollins, Harper Perennial.
[151] Ibid.
[152] "Sage Against the Machine," Tunku Varadarajan, September 1-2, 2018, Wall Street Journal.
[153] "5 high-Profile Cryptocurrency hacks," Lorraine Ryshin, October 1, 2018, Ibinex News.com.
[154] "Facebook Hackers Access Nearly 50 Million Accounts," Deepa Seetharaman and Robert McMillan, September 29, 2018, Wall Street Journal.
[155] "Driverless-Car Legislation is Unsafe at This Speed," Ralph Hader, August 23, 2018, Wall Street Journal.
[156] "US Power grid needs defense against looming cyberattacks," Melanie Kenderdine and David Jermain, March 23, 2018, The Hill.
[157] Ibid.
[158] "America Goes on the Cyberoffensive," Dave Weinstein, August 29, 2018, Wall Street Journal.
[159] Ibid.
[160] "Algorithms With Minds of Their Own," Curt Levy and Ryan Hagemann, Wall Street Journal, November 13, 2017.
[161] "The Impact of the Internet on Society: A Global Perspective," Manuel Castellas, September 8, 2014, Technology Review.com (Chair, Professor of Communication Technology at the University of Southern California, Los Angeles).
[162] "For Teens, Face Time Loses Out To Screens," Betsy Morris, September 11, 2018, Wall Street Journal.
[163] "The Impact of the Internet on Society: A Global Perspective," Manuel Castells, September 8, 2014, Technology Review.com (Chair, Professor of Communication Technology at the University of Southern California, Los Angeles).
[164] Information Technology and Moral Values," Stanford Encyclopedia of Philosophy, June 12, 2012.
[165] "What Happens When Artificial intelligence Turns On Us?", Erica R. Hendry, January 21, 2014, Smithsonian.com.
[166] Ibid.

[167] "Sage Against the Machine," Tunku Varadarajan, September 1-2, 2018, Wall Street Journal.

[168] "Technology isn't just changing society—it's changing what it means to be human: A conversation with historian of science Michael Bess," Sean Illing, February 23, 2018, Vox.com.

[169] Ibid.

[170] Ibid.

[171] Ibid.

[172] Information Technology and Moral Values," Stanford Encyclopedia of Philosophy, June 12, 2012

[173] Ibid.

[174] "On the Intrinsic Value of Information Objects and the Infosphere," L. Floridi, 2003, Ethics and Information Technology, 4(4): 287-304.

[175] Artificial intelligence—can we keep it in the box?", Huw Price and Jean Tallinn, August 5, 2014, The Conversation.com.

[176] "Philosophical Plumbing," Mary Midgley, Royal Institute of Philosophy Supplement 33:139-151 (1992).

[177] Information Technology and Moral Values," Stanford Encyclopedia of Philosophy, June 12, 2012.

[178] Ibid.

[179] "Artificial intelligence—can we keep it in the box?", Huw Price and Jean Tallinn, The Conversation.com.

[180] Ibid.

[181] "Stephen Hawking: Artificial Intelligence Could Wipe Out Humanity When it Gets Too Clever as Humans Will be Like Ants," Andrew Griffin, October 8, 2015, Independent.com.

[182] "What Happens When Artificial intelligence Turns On Us?", Erica R. Hendry, January 21, 2014, Smithsonian.com.

[183] "Artificial intelligence—can we keep it in the box?", Huw Price and Jean Tallinn, The Conversation.com.

[184] Ibid.

[185] "The Relentless Pace of Automation," David Rotman, February 13, 2017, Technology Review.

[186] "The Human Promise of the Revolution," Kai-Fu Lee, September 15-16, Wall Street Journal.

[187] Dr. Lee is chairman and CEO of Sinovation Ventures and former

president of Google China. This referenced essay is adapted from Dr. Lee's new book: "AI Superpowers: China, Silicon Valley and the New World Order," Houghton Mifflin Harcourt.

[188] "The Human Promise of the Revolution," Kai-Fu Lee, September 15-16, Wall Street Journal.

[189] "The ethics of Artificial Intelligence," Justin Lee, June 26, 2018, GrowthBot.org.

[190] "Pence Cautions Google on China," Michael C. Bender and Dustin Volz, October 5, 2018. Wall Street Journal.

[191] "Beijing Expands Its Cybersecurity Regulations," Shan Li, October 6-7, 2018, Wall Street Journal.

[192] Ibid.

[193] "The Big Hack: How China Used a Tiny Chip to Infiltrate U.S. Companies," October 4, 2018, Bloomberg Businessweek.

[194] Ibid.

[195] "Exclusive: FBI probes FDIC hack linked to China's military—sources," Dustin Volz and jason Lang, December 23, 2016, Reuters World News.

[196] "The Big Hack: How China Used a Tiny Chip to Infiltrate U.S. Companies," October 4, 2018, Bloomberg Businessweek.

[197] "New poll of rural Americans shows deep cultural divide with urban centers," Jose A. Delreal and Scott Clement, June 17, 2017, The Washington Post.

[198] "Government Can't Rescue the Poor," Phil Gramm and John E. Early, October 11, Wall Street Journal.

[199] "The Differences Between City, Suburban, and Rural Living," February 2, 2018, Property Management, Inc., https://www.rentmi.com.

[200] "The Differences Between City, Suburban, and Rural Living," February 2, 2018, Property Management, Inc., https://www.rentmi.com.

[201] "The Relentless Pace of Automation," David Rotman, February 13, 2017, MIT Technology Review.

[202] "The Human Promise of the Revolution," Kai-Fu Lee, September 15-16, Wall Street Journal.

[203] Ibid.

[204] Ibid.

[205] "The Relentless Pace of Automation," David Rotman, February 13, 2017, MIT Technology Review.

[206] "The Human Promise of the Revolution," Kai-Fu Lee, September 15-16, Wall Street Journal.

[207] "Explore our featured insights," McKinsey & Company, https://www.bing.com/search?q=McKinsey+%26+Company+Global+Institute%2C+in+2008%2C+Microsoft+was+the+only+such+tech+company+that+made+it+into+the+top+ten+largest+companies+globally.+Apple+came+in+next+at+39%2C+and+Google+at+51.+By+2018%2C+the+top+five+spots+were+claimed+by+the+top+five+tech+giants%2C+both+in+the+US+and+globally.&qs=n&form=QBLH&sp=-1&pq=&sc=8-0&sk=&cvid=33BAB87D3B1A495AA52A8DA011DEF8F7.

[208] "The Relentless Pace of Automation," David Rotman, February 13, 2017, MIT Technology Review.

[209] "The Relentless Pace of Automation," David Rotman, February 13, 2017, MIT Technology Review.

[210] "Government Can't Rescue the Poor," Phil Gramm and John F. Early, October 11, 2018, Wall Street Journal.

[211] Ibid.

[212] "Stephen Hawking: Artificial Intelligence Could Wipe Out Humanity When it Gets Too Clever as Humans Will be Like Ants," Andrew Griffin, October 8, 2015, Independent.com.